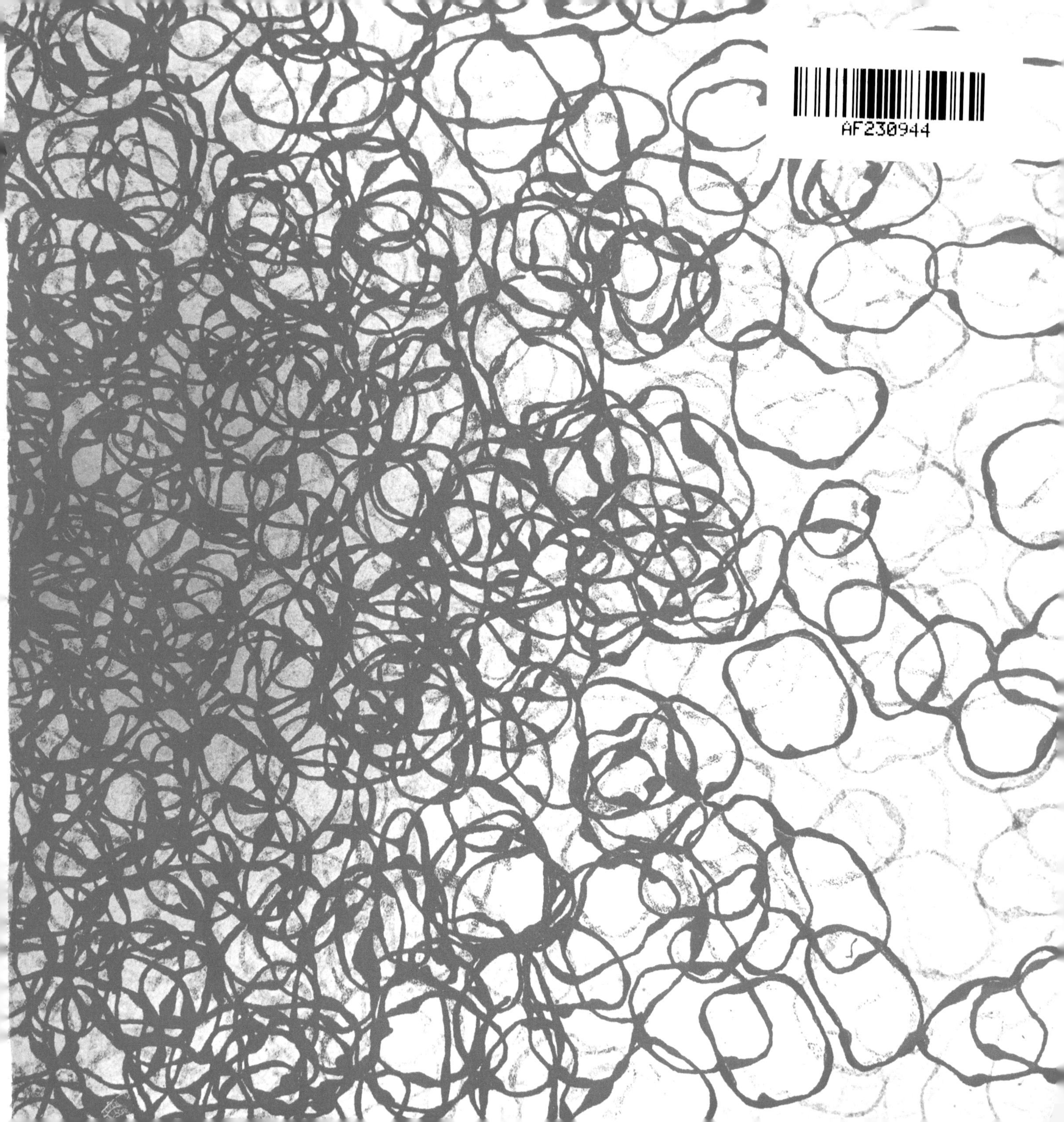
AF230944

Origin story by Don Drake

Artwork by Kent Manske

and change became

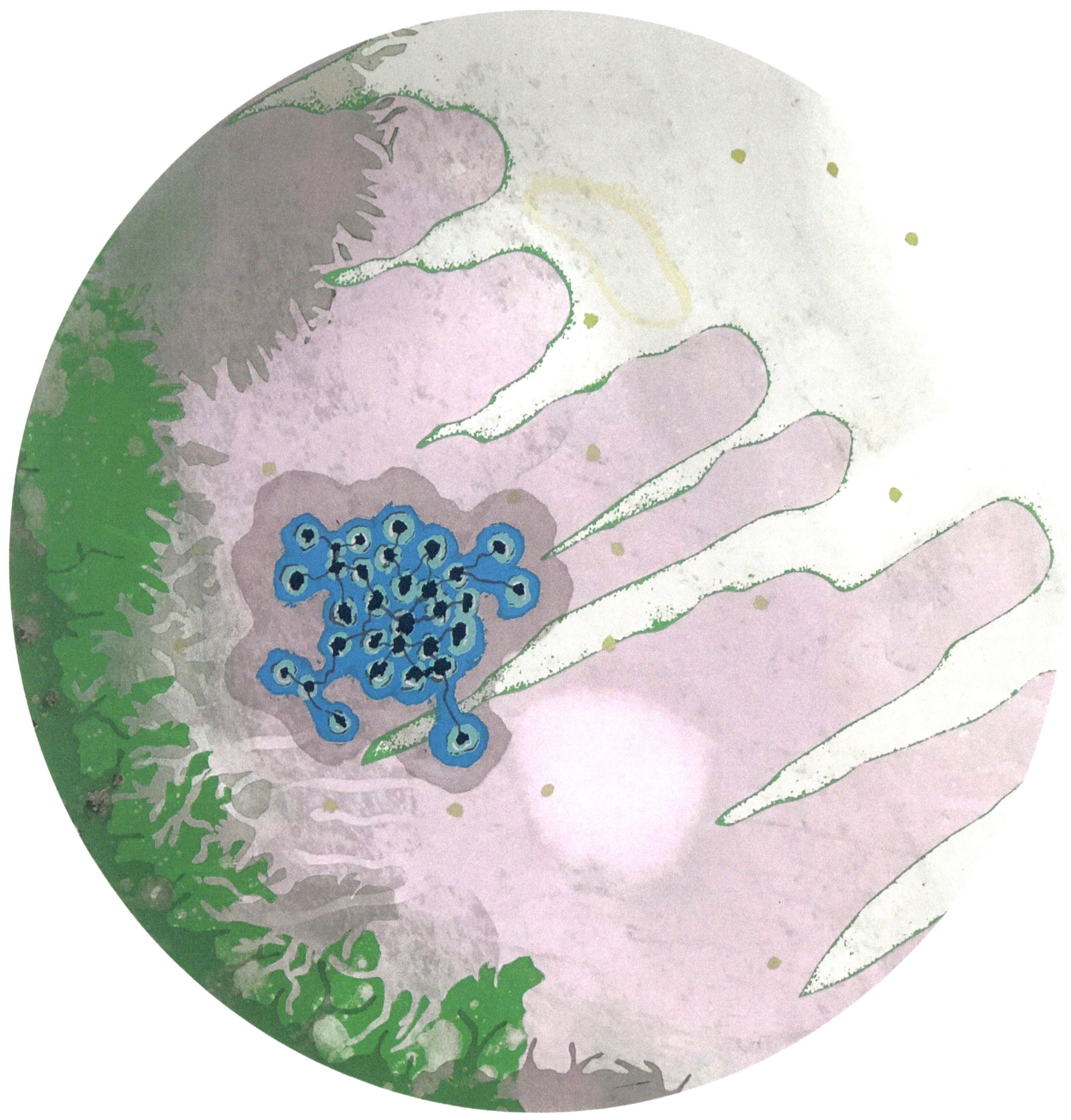

They formed a circle
around the fire
and she looked at their gathering.

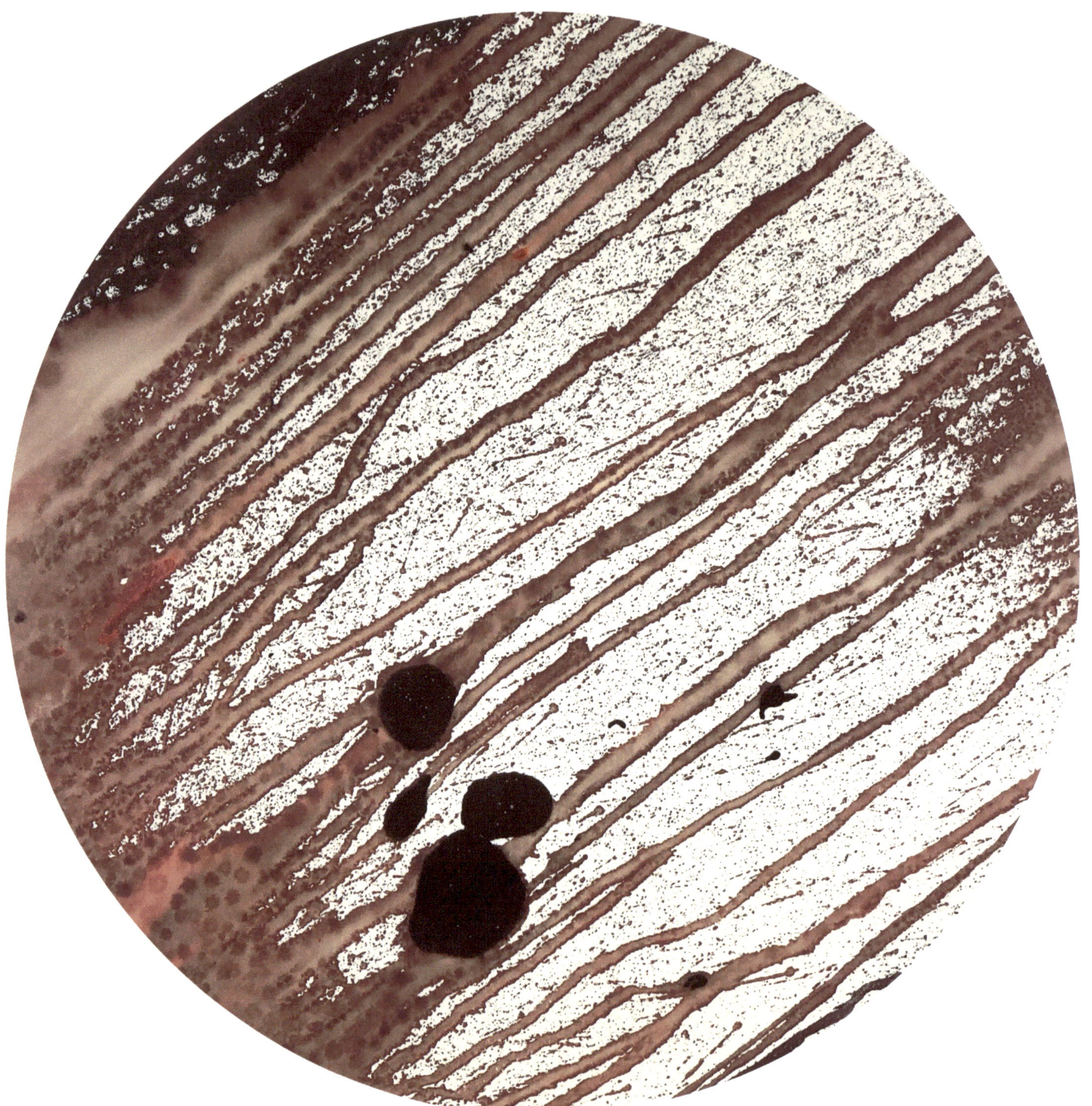

The group quieted
and eyes vectored her direction.

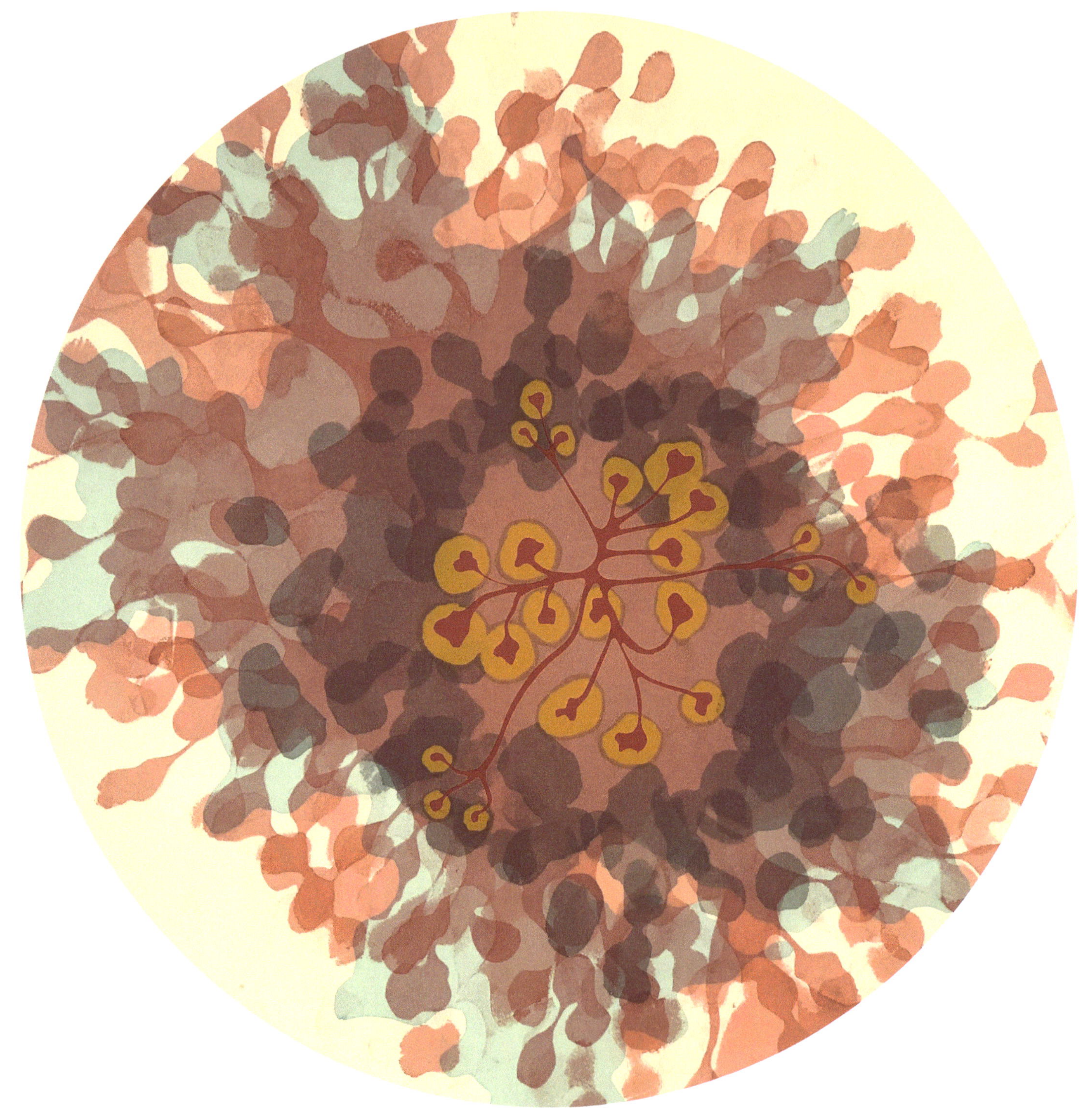

Silent attention coalesced.

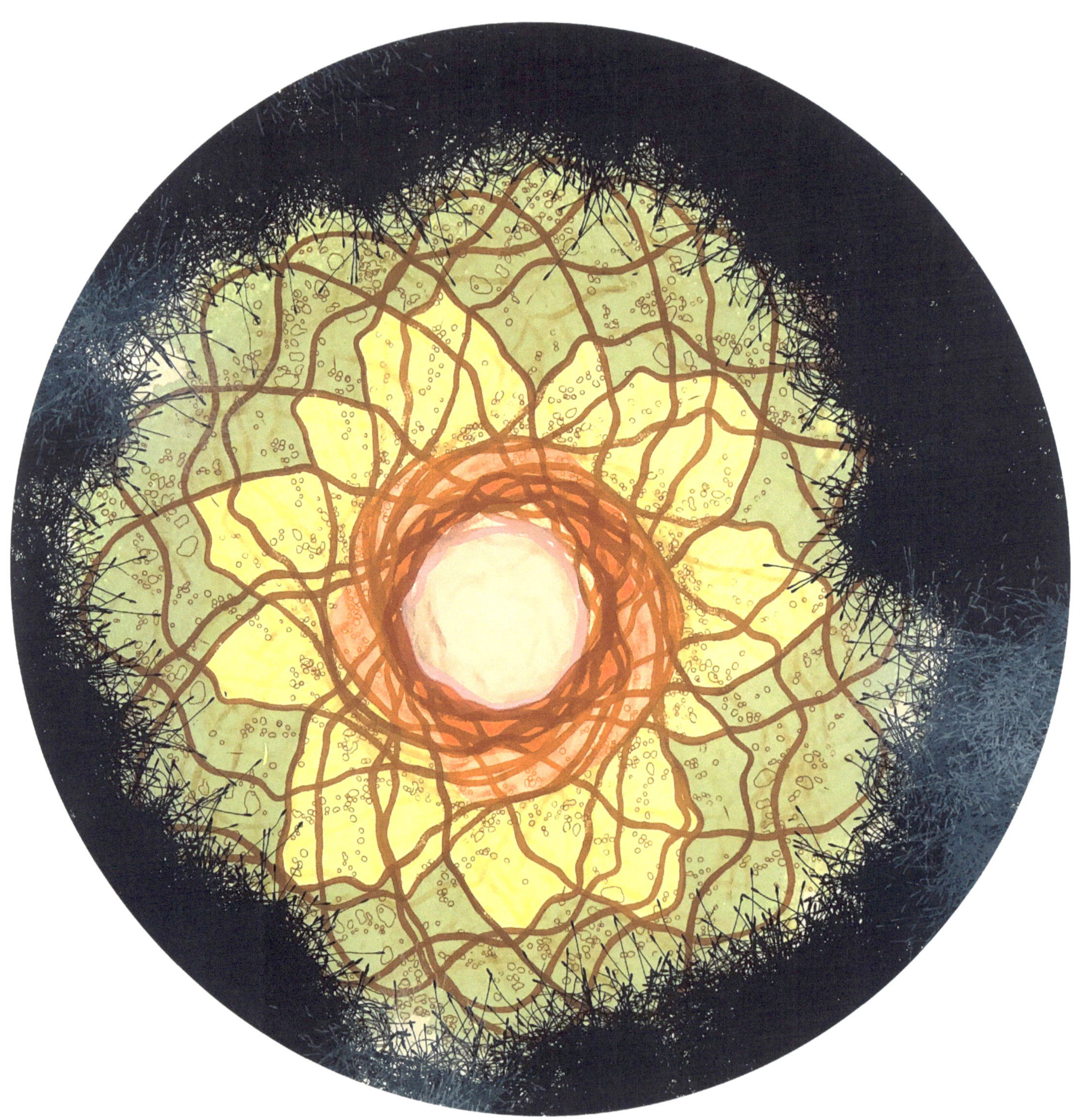

Tonight
I tell the story of a beginning.

Of the instant Existence
tried for perfection.

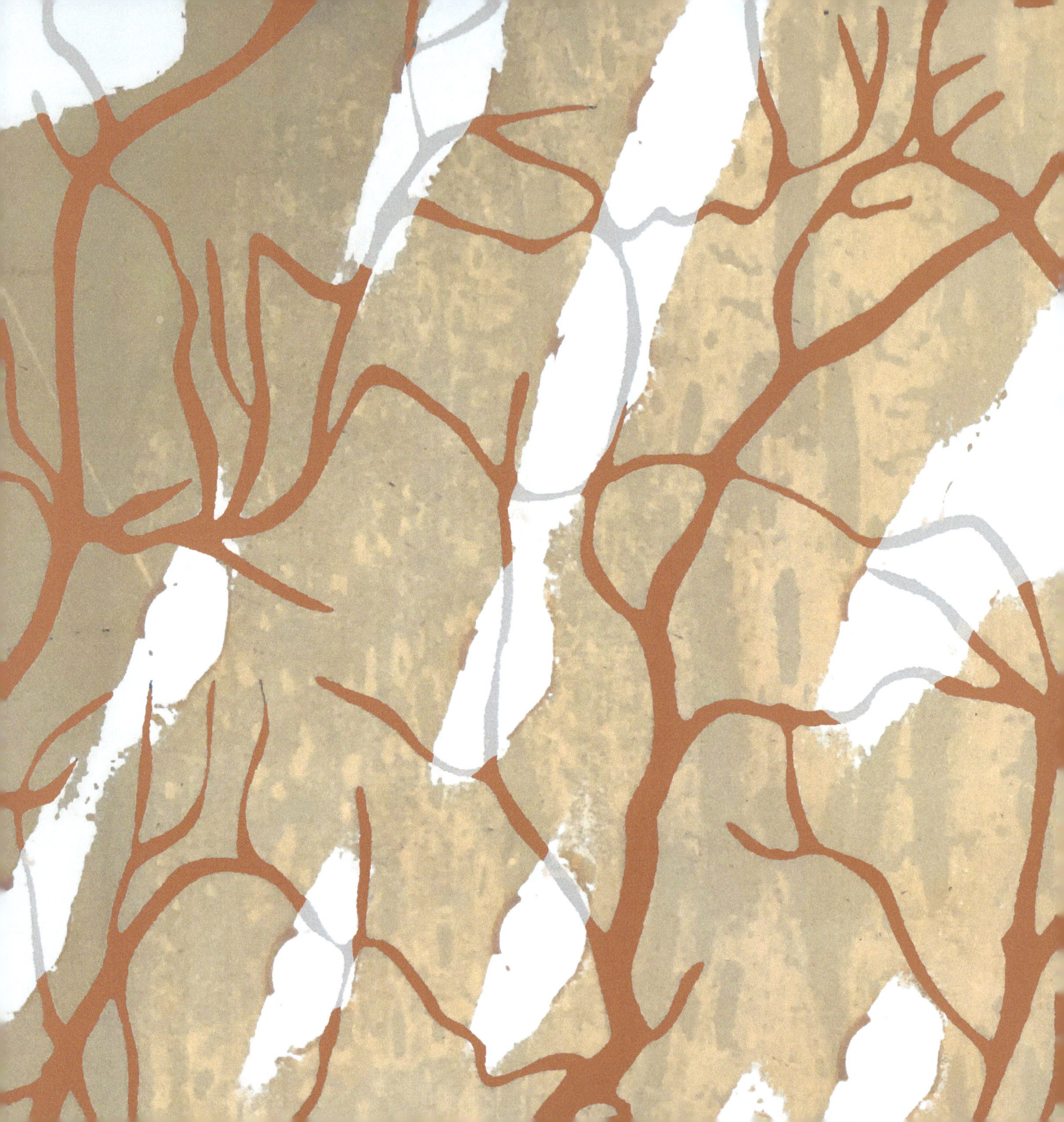

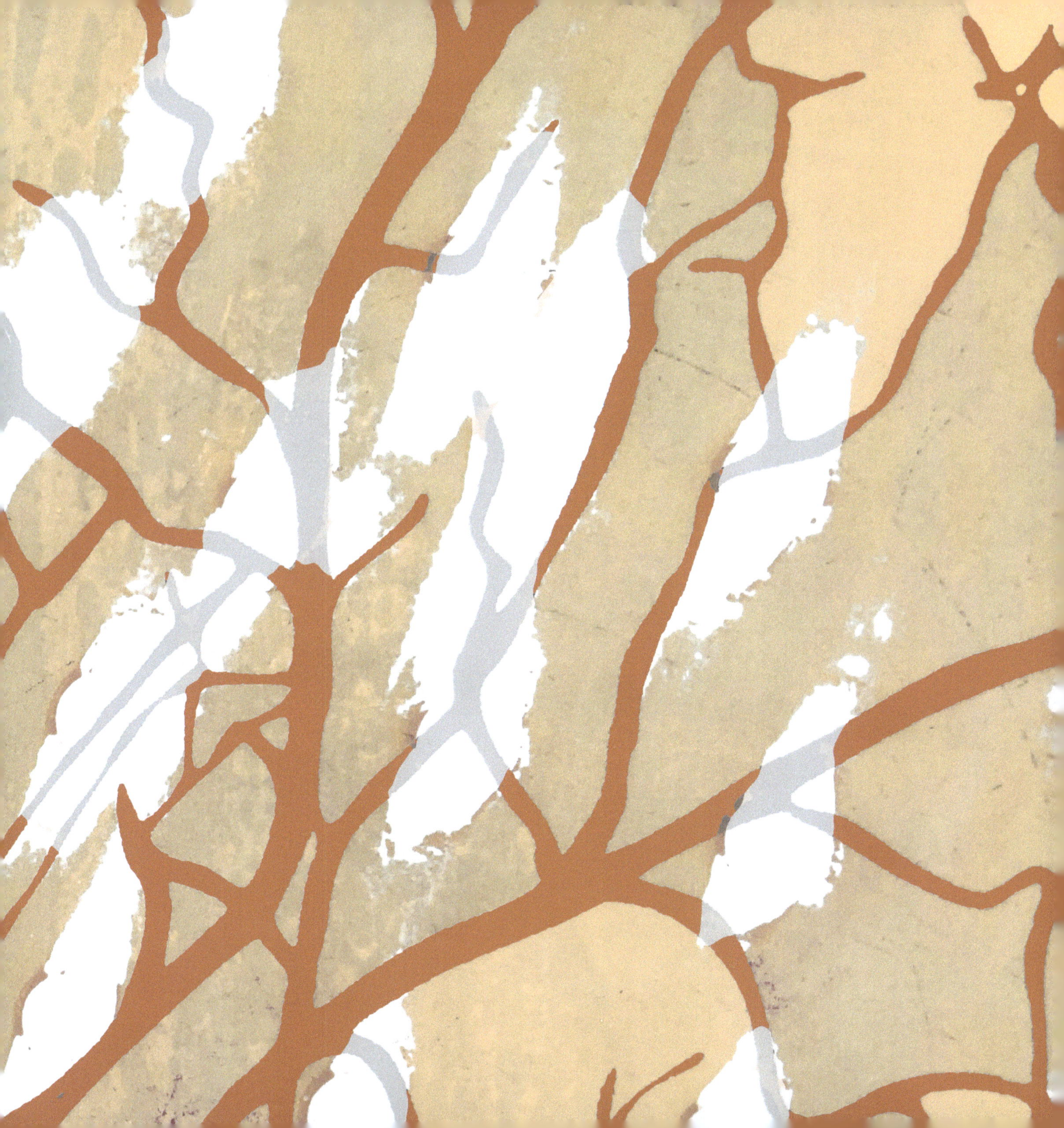

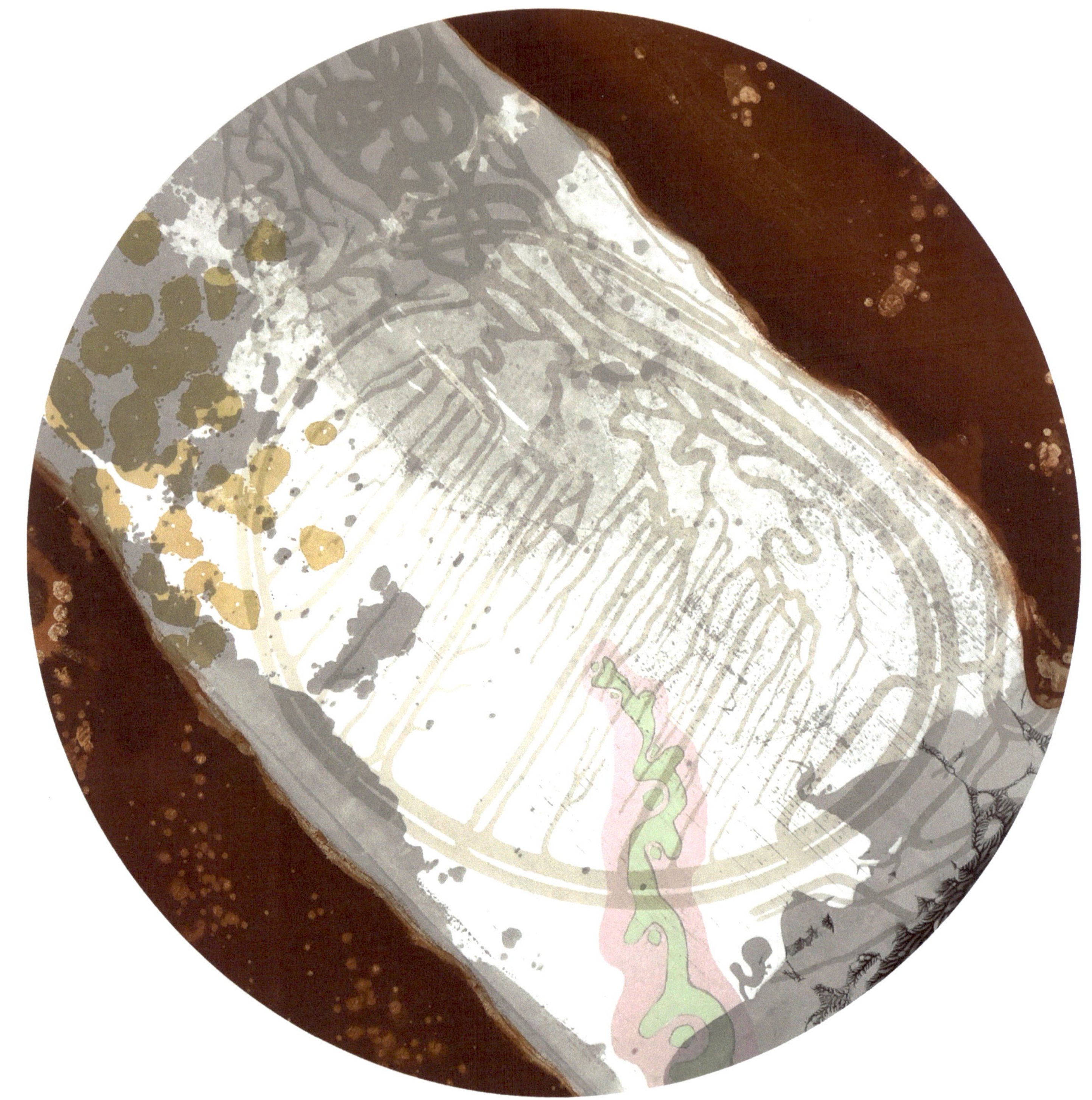

At a place before time,
in a time before space,
Existence was a unity.

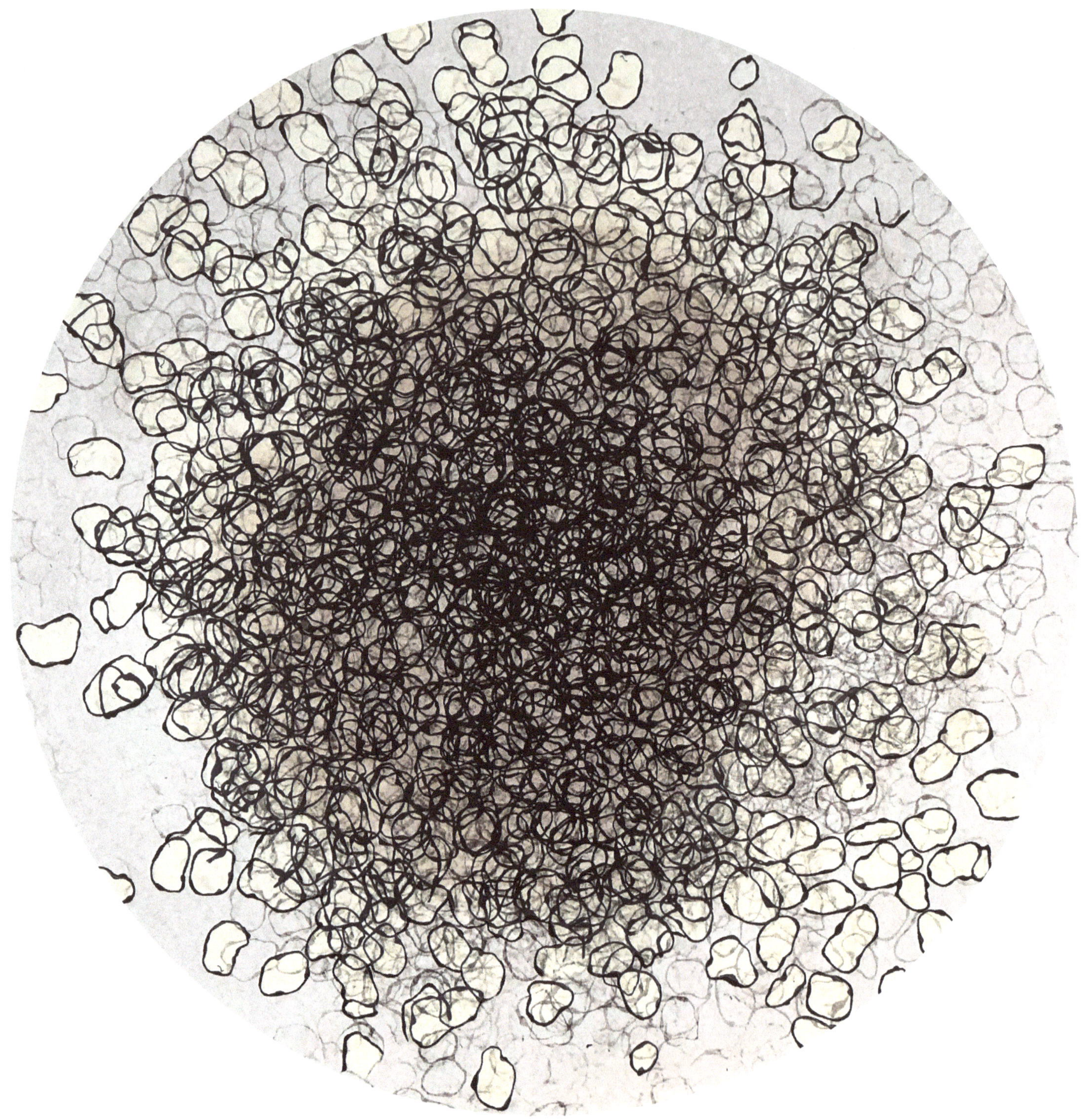

The unity of energy, matter and dimension.

The unity of cause and effect.

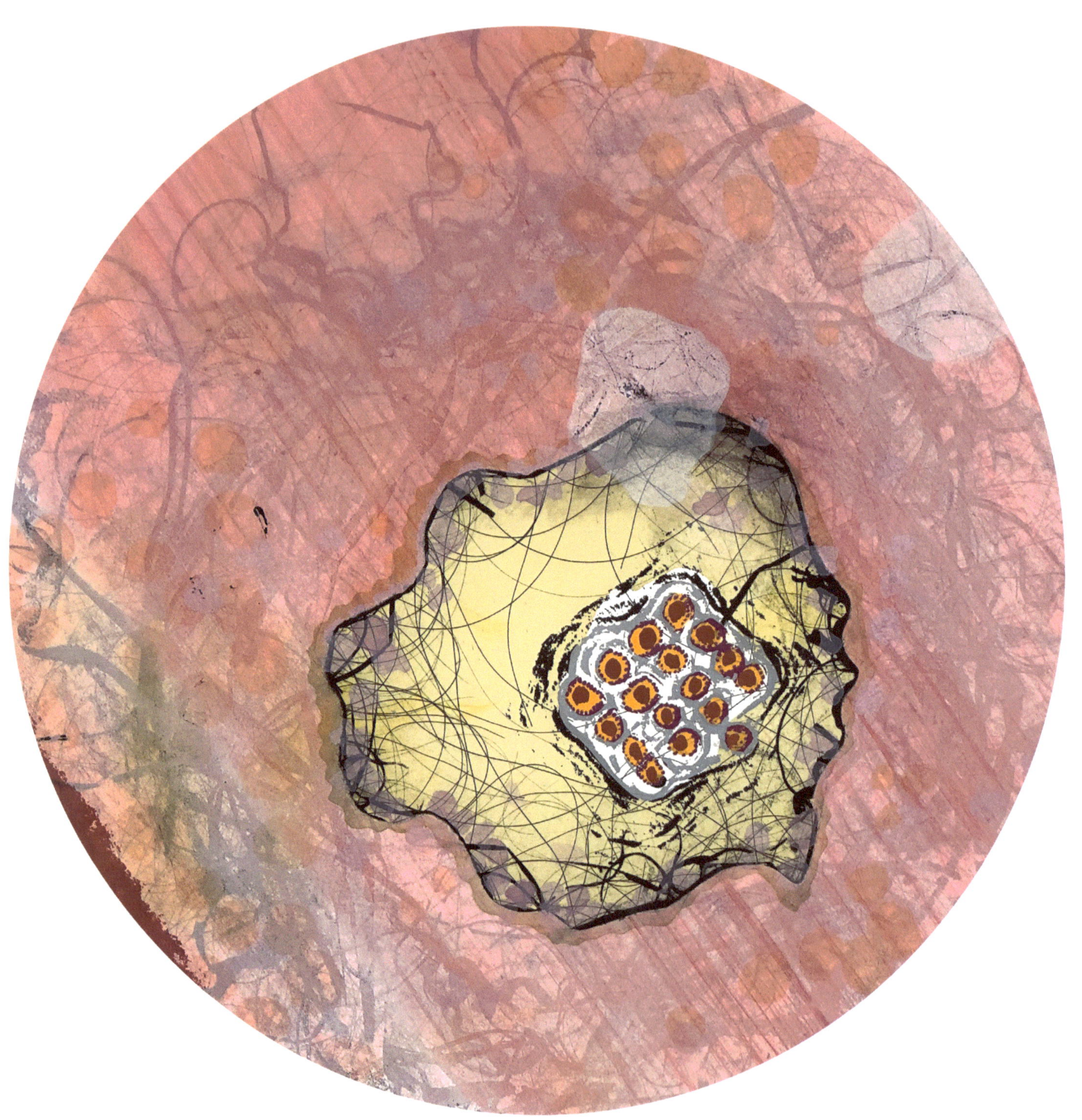

You and I could not be there
because there was no 'there'
and no 'you' or 'I'.

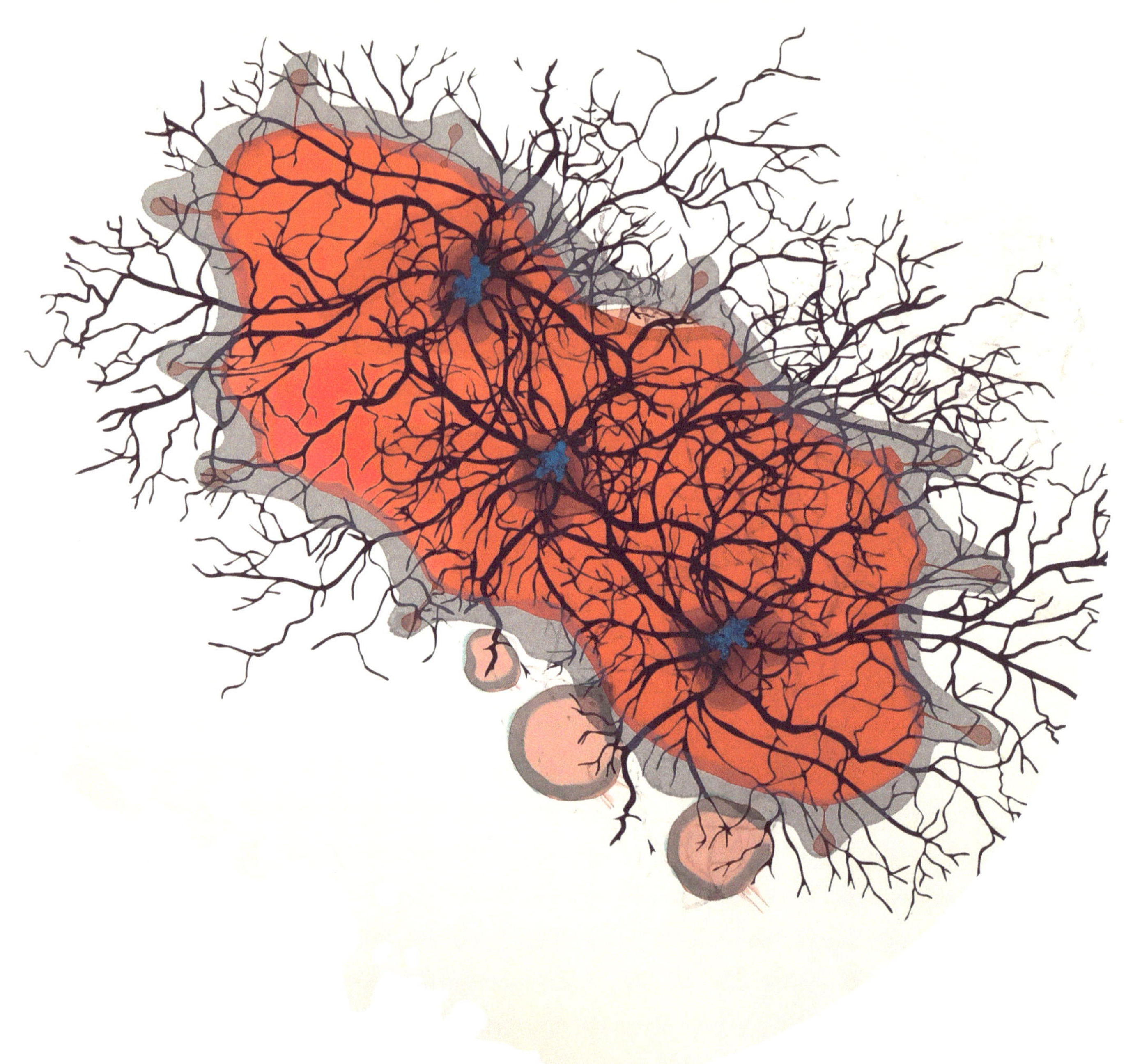

If we could have looked,
we would have found all space
and all time
in one place
at one instant.

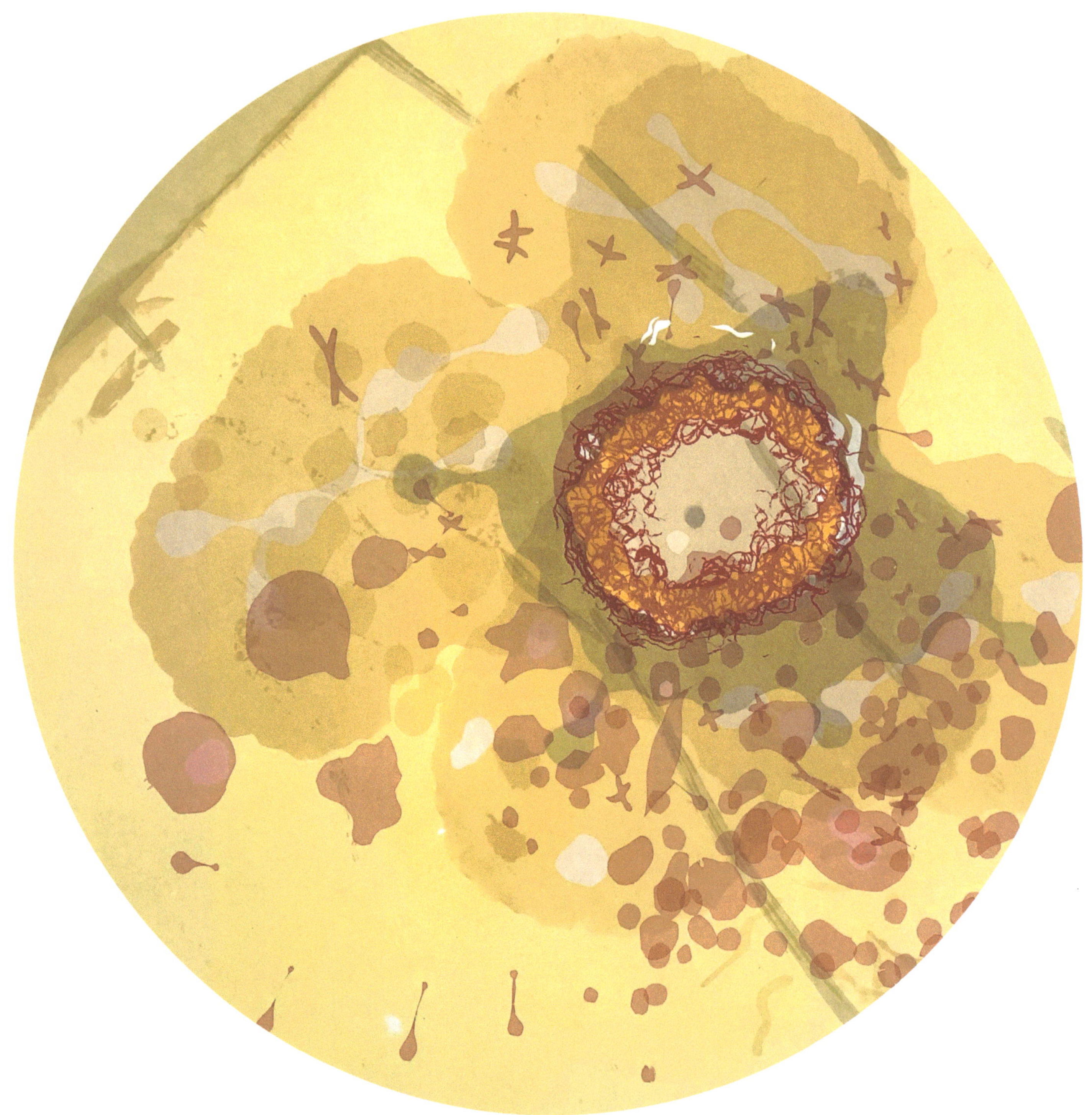

All matter
existing only as a unified potential,
unvarying at a timeless point.

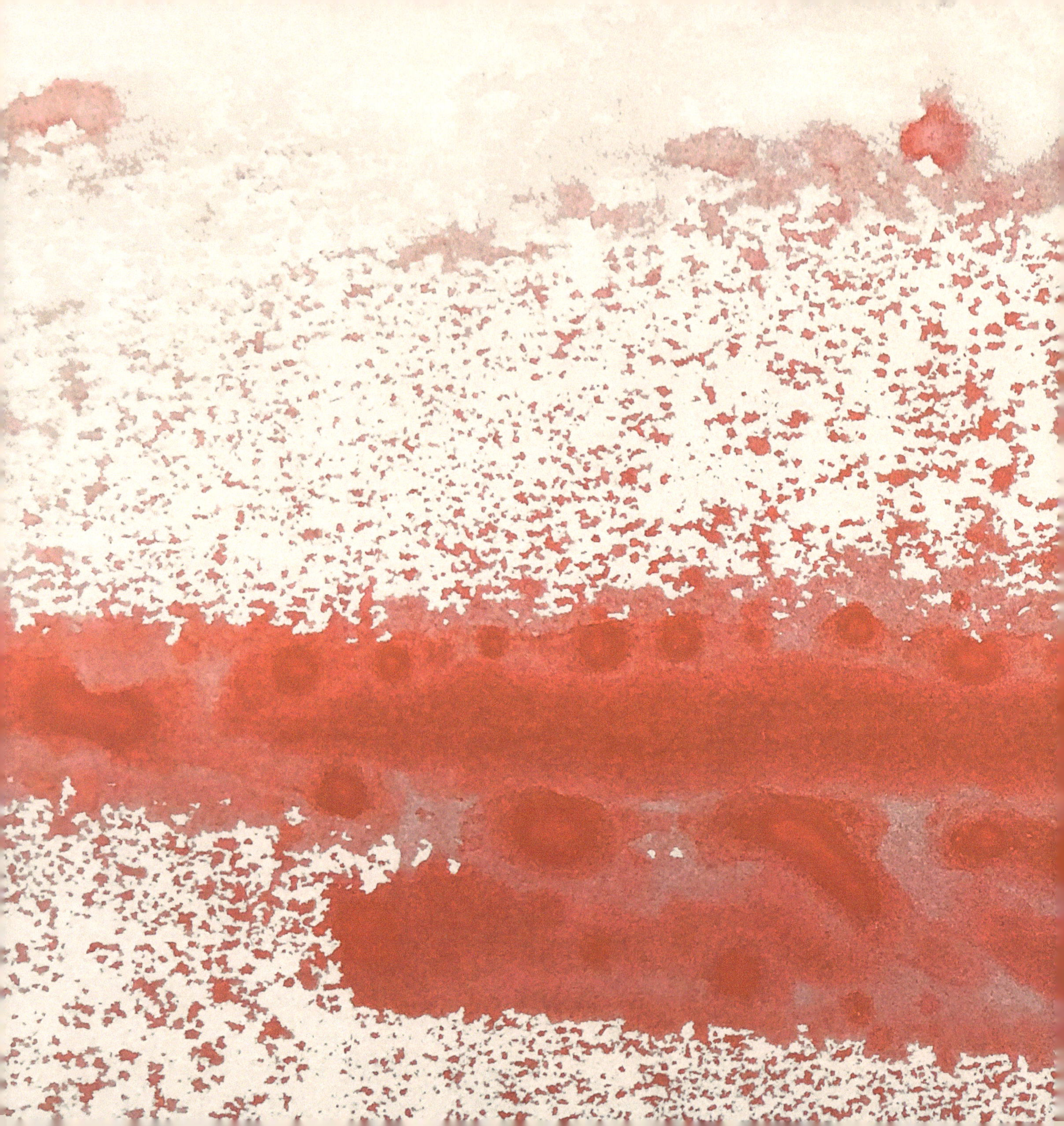

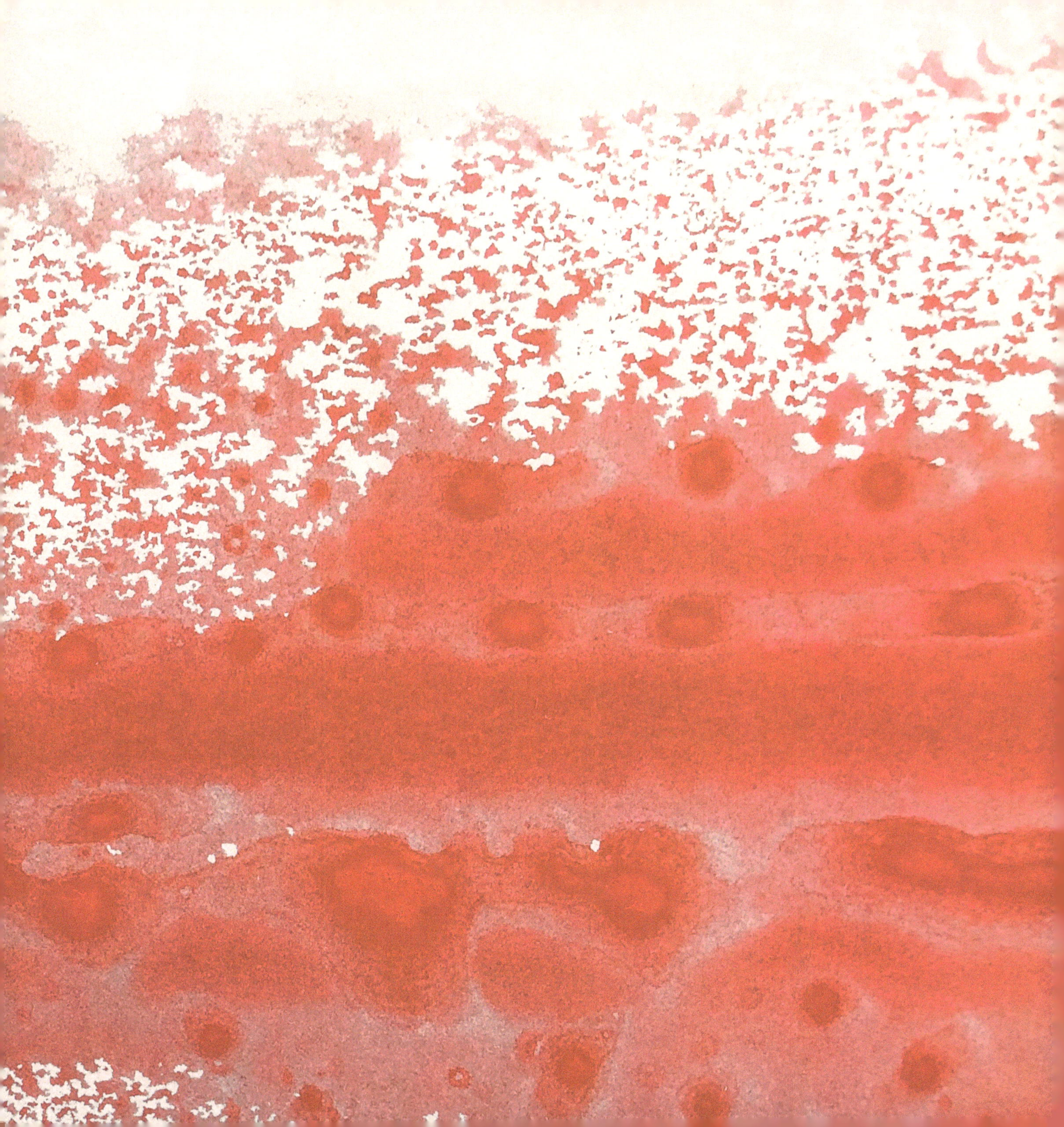

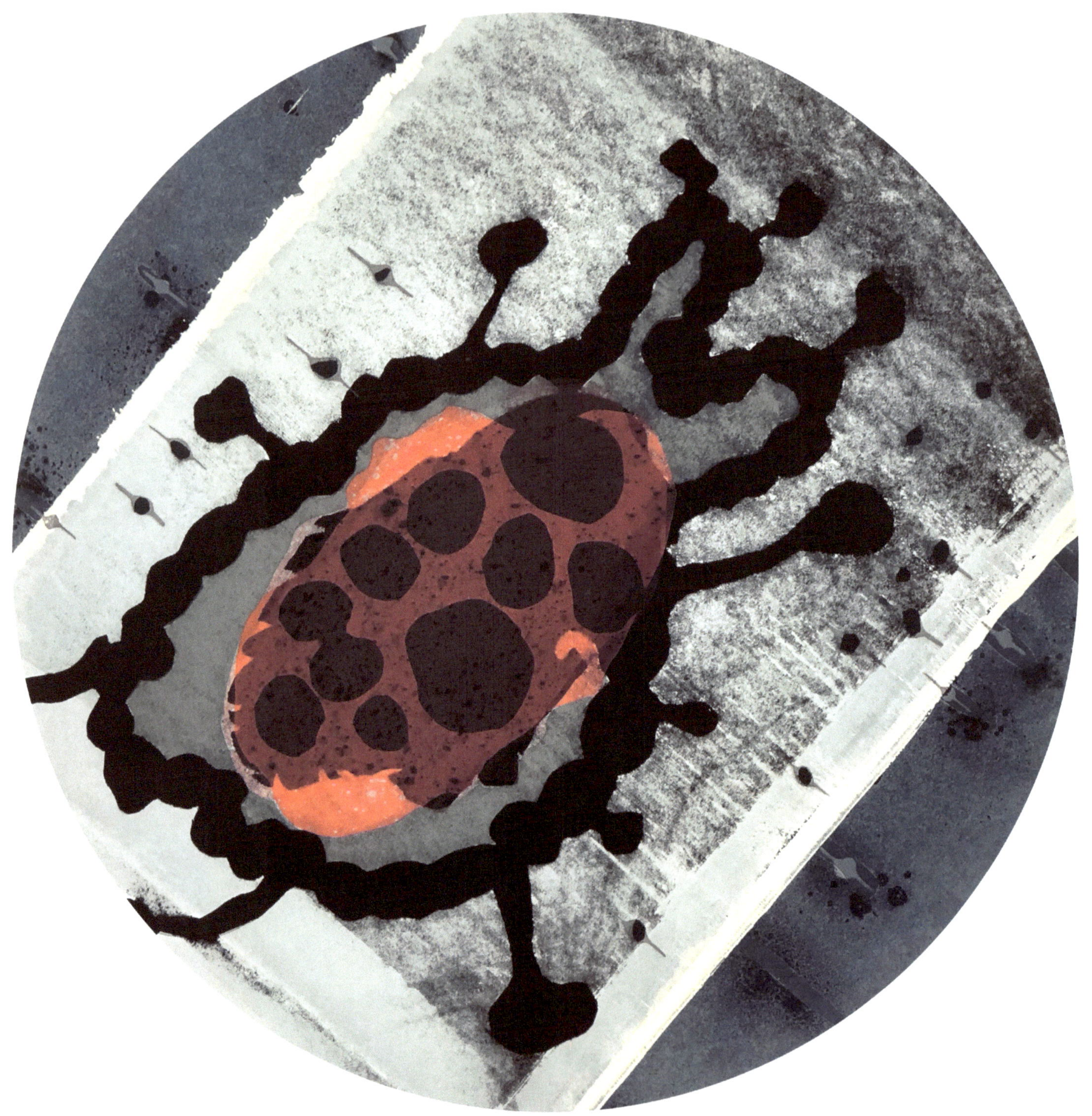

But Existence believed
there was a shadow on its unity,
for it had an unrealized companion;
Entropy.

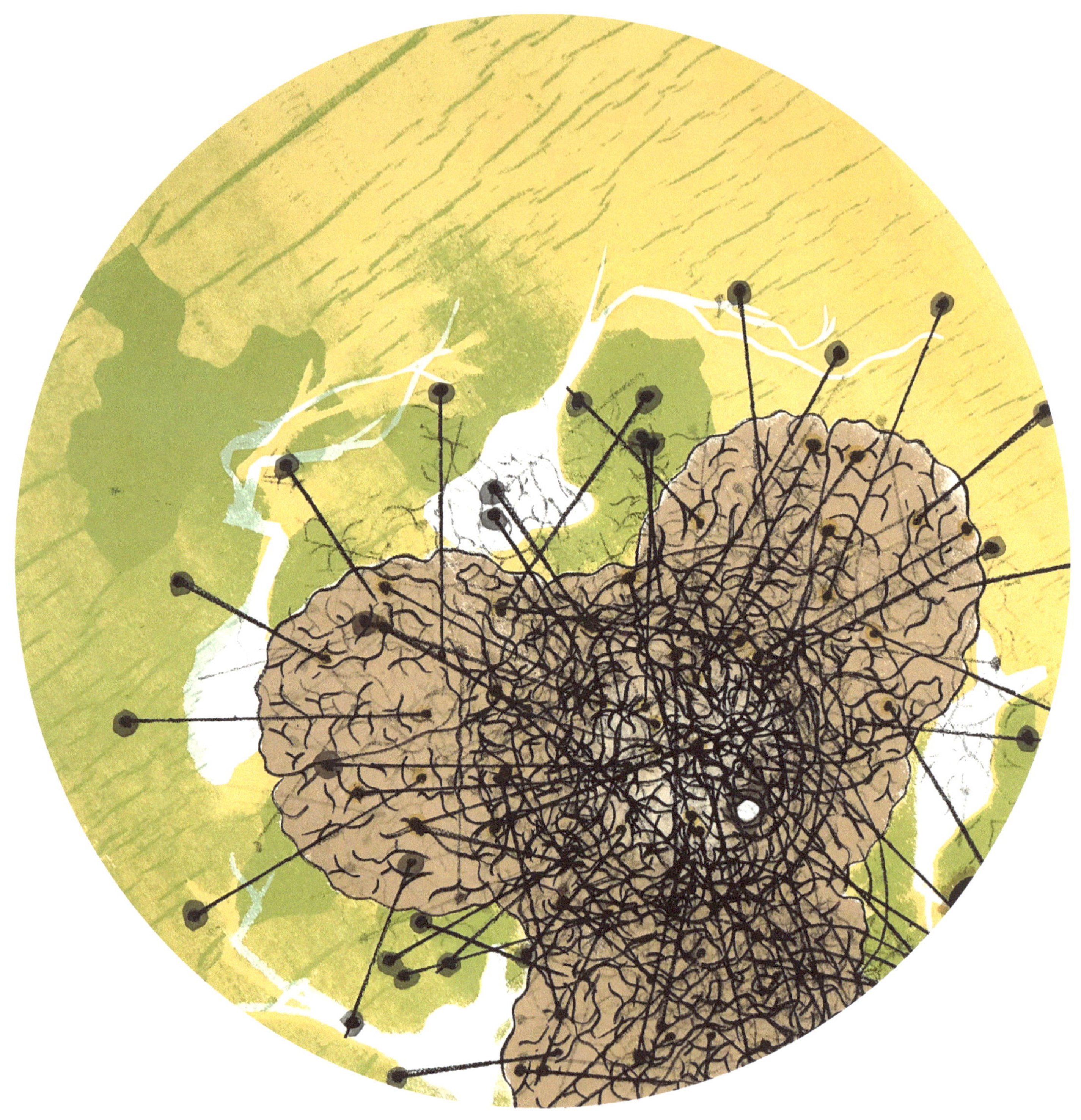

Entropy was the harbinger
of difference, variety, and change.

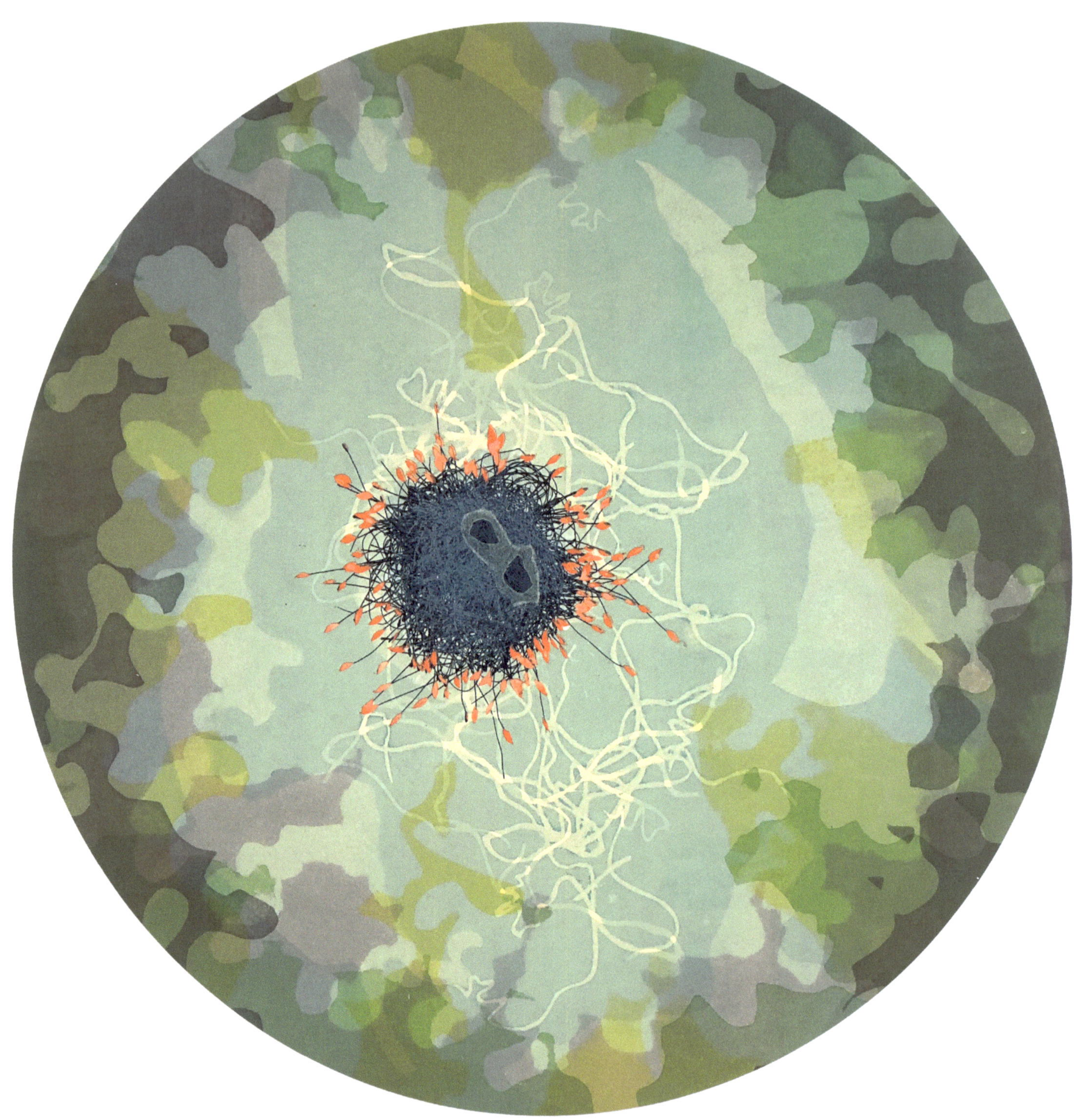

However, in a unity
there is not two of anything,
so Entropy was only an idea
crushed to zero
by the nature of Existence.

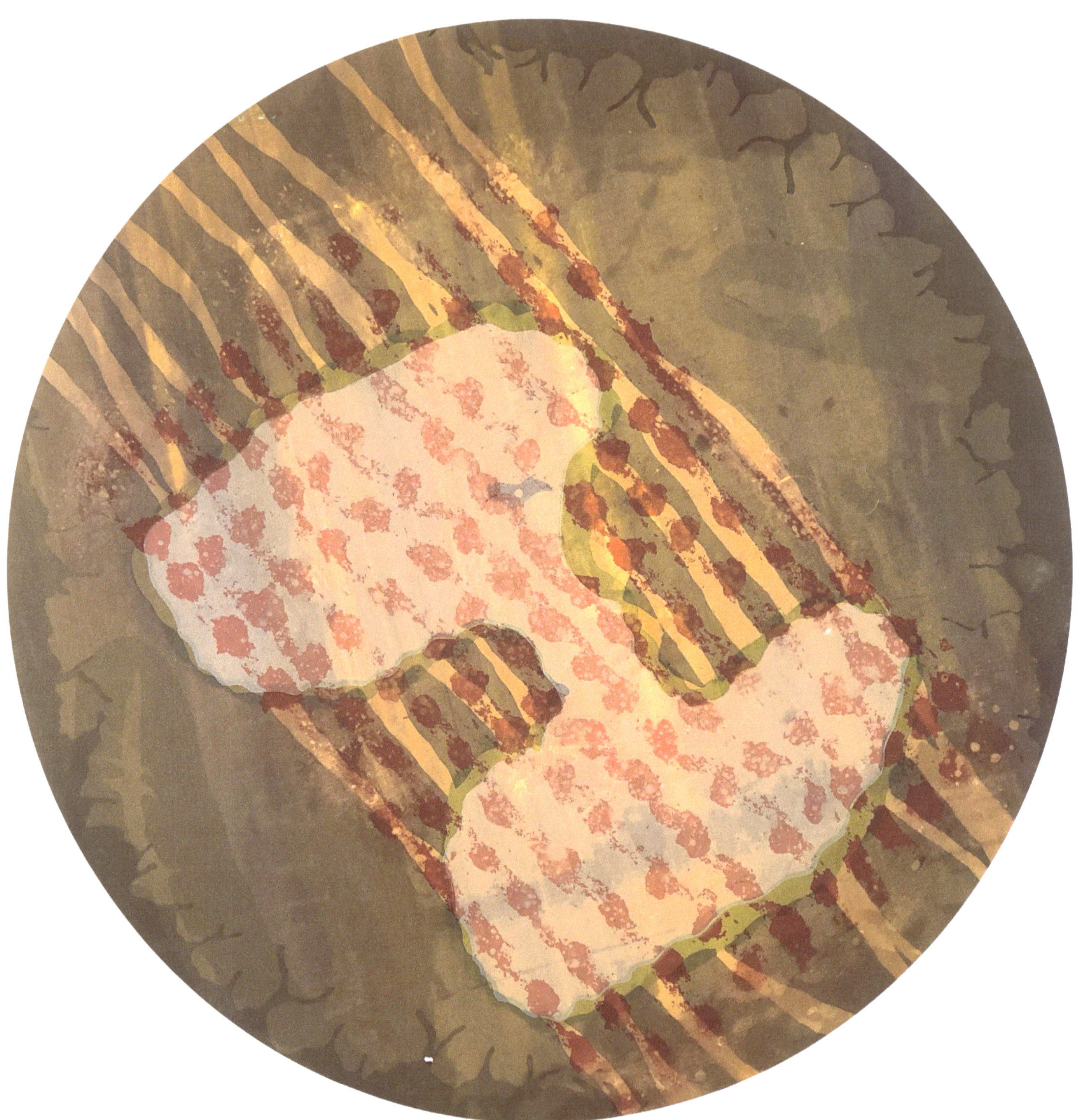

Still, zero is not one,
so did Existence truly embody unity?

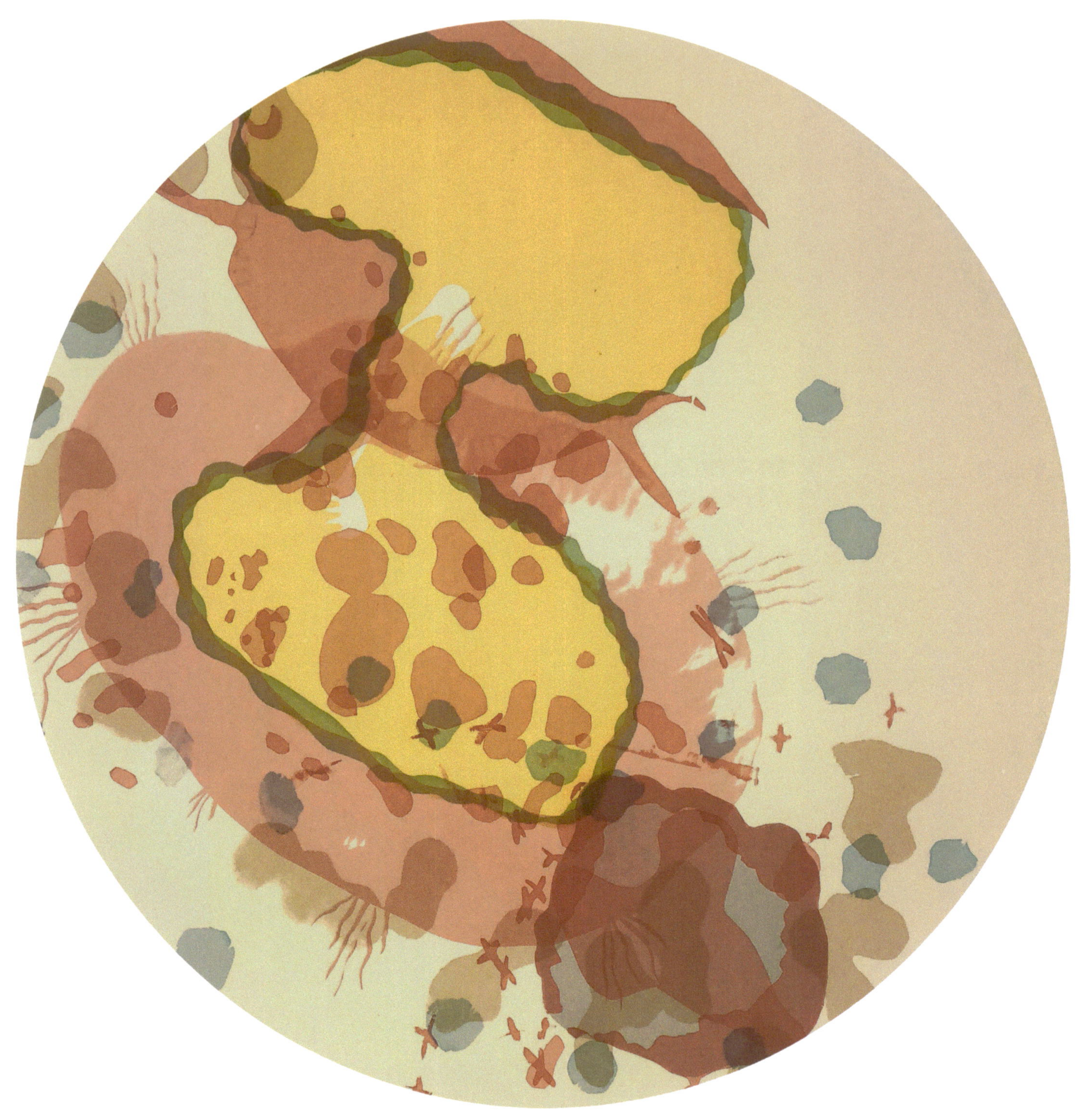

The question plagued Existence
because, though it was the unification
of space, time and energy,
it believed it was not perfect
while this outsider
could be imagined.

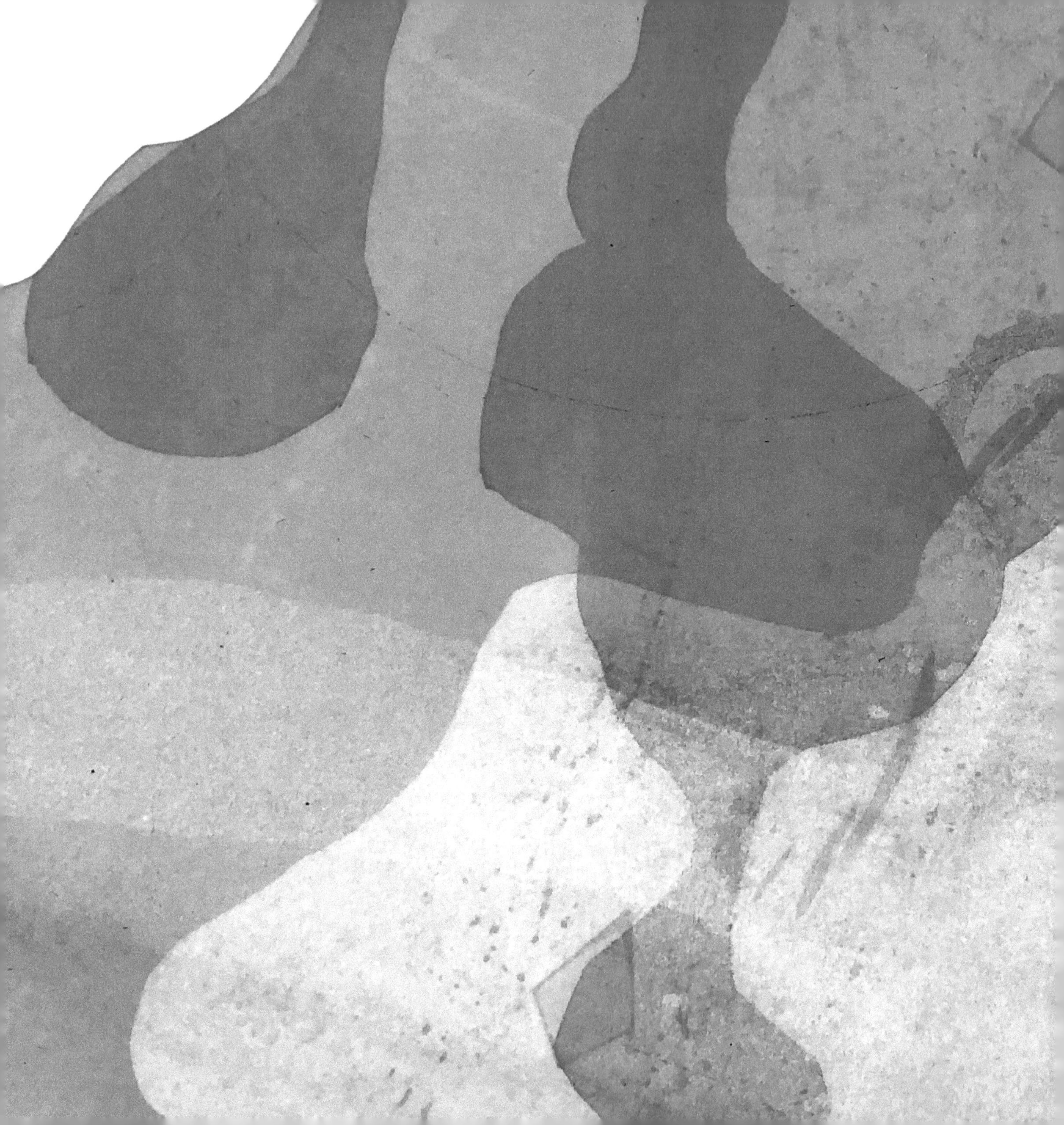

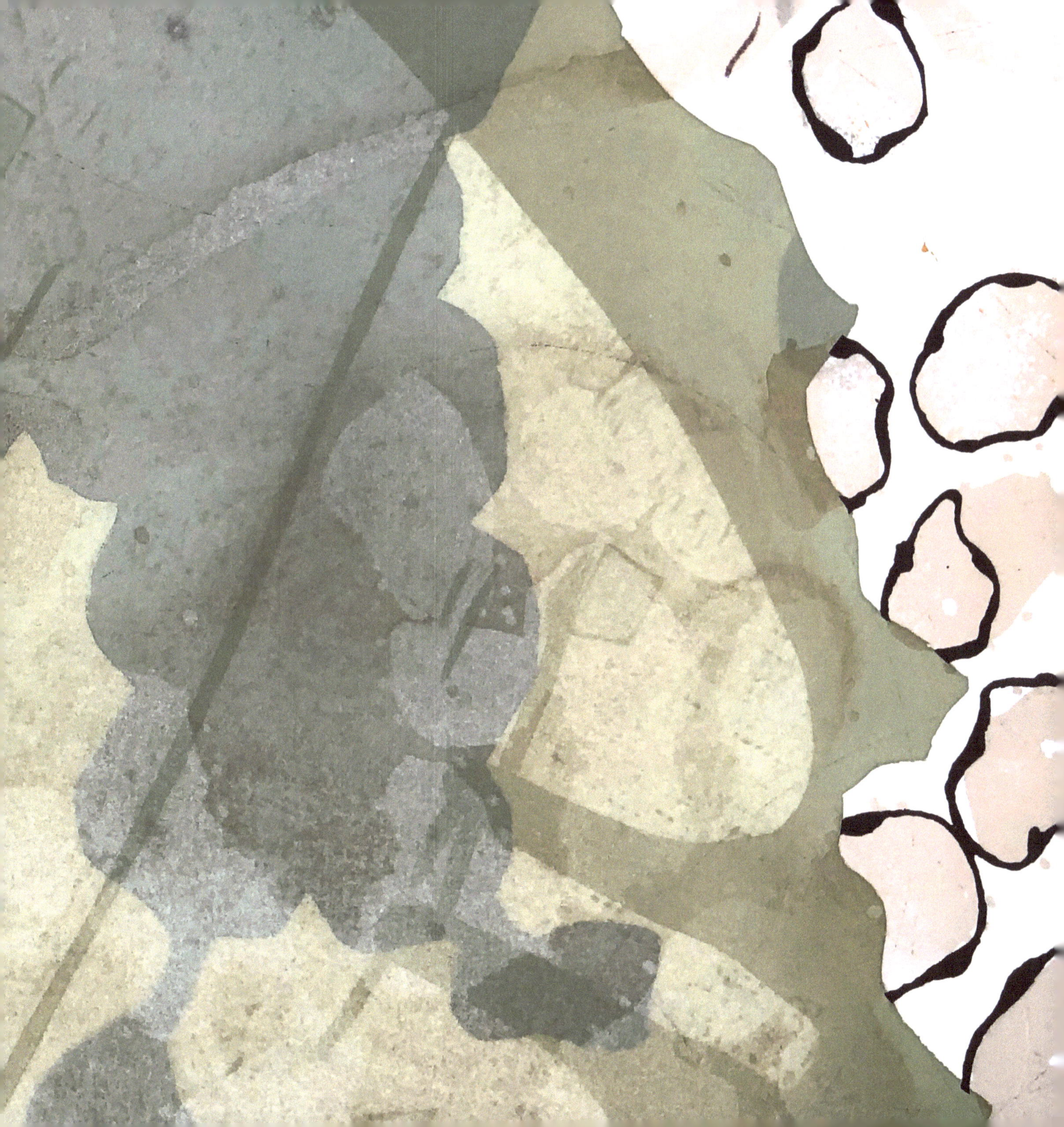

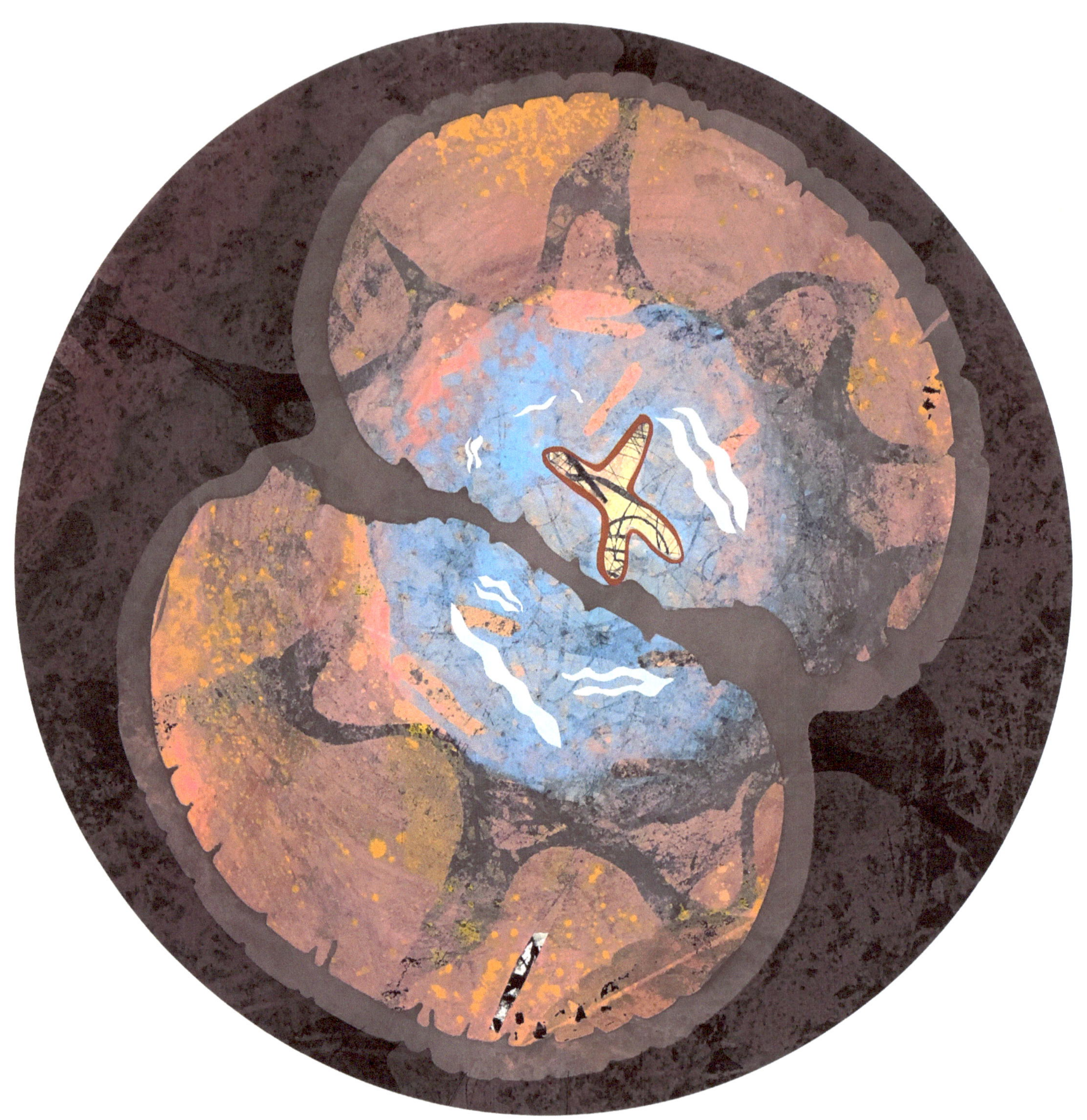

If Existence could enfold Entropy
it could truly be the perfect unity.

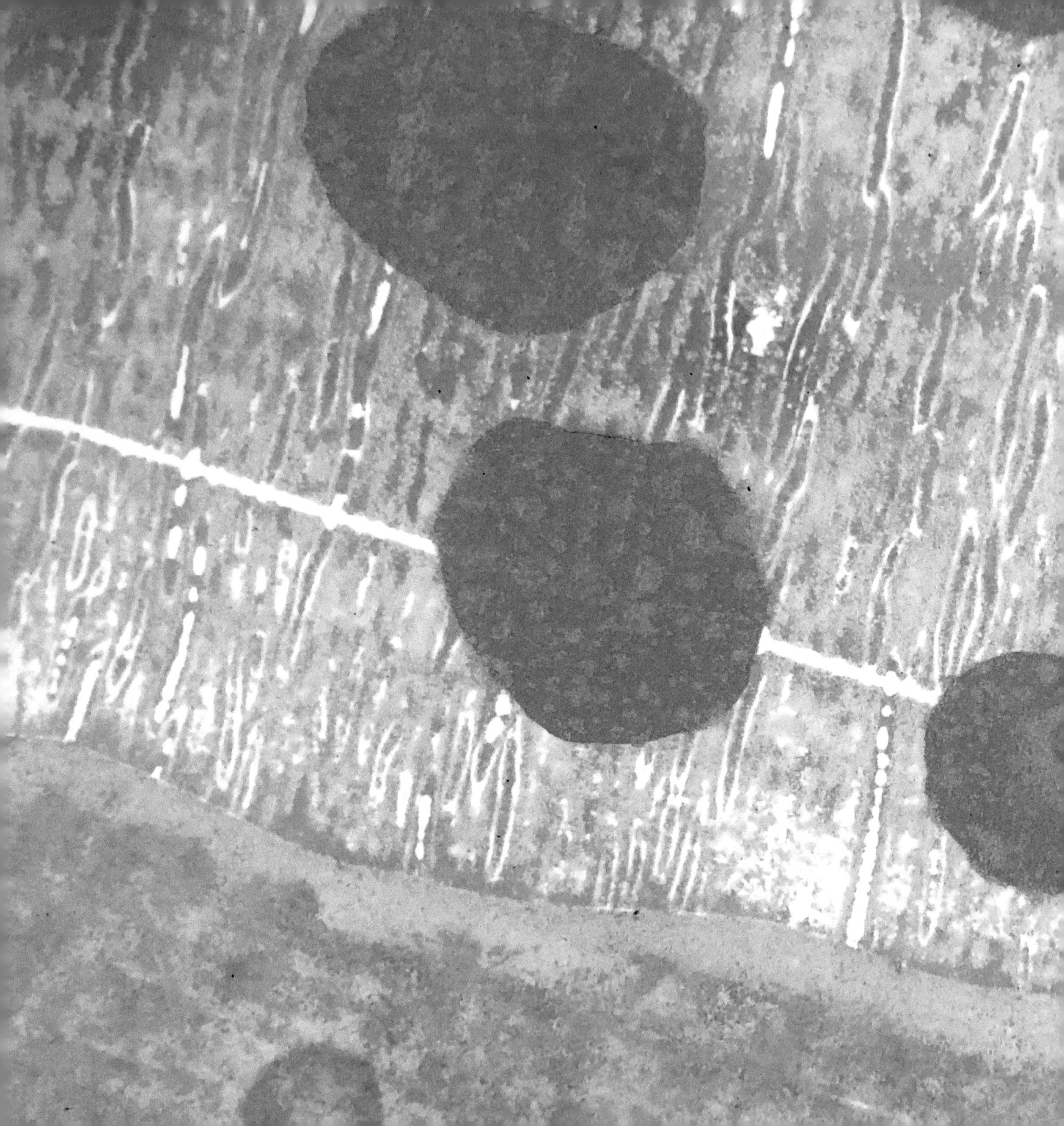

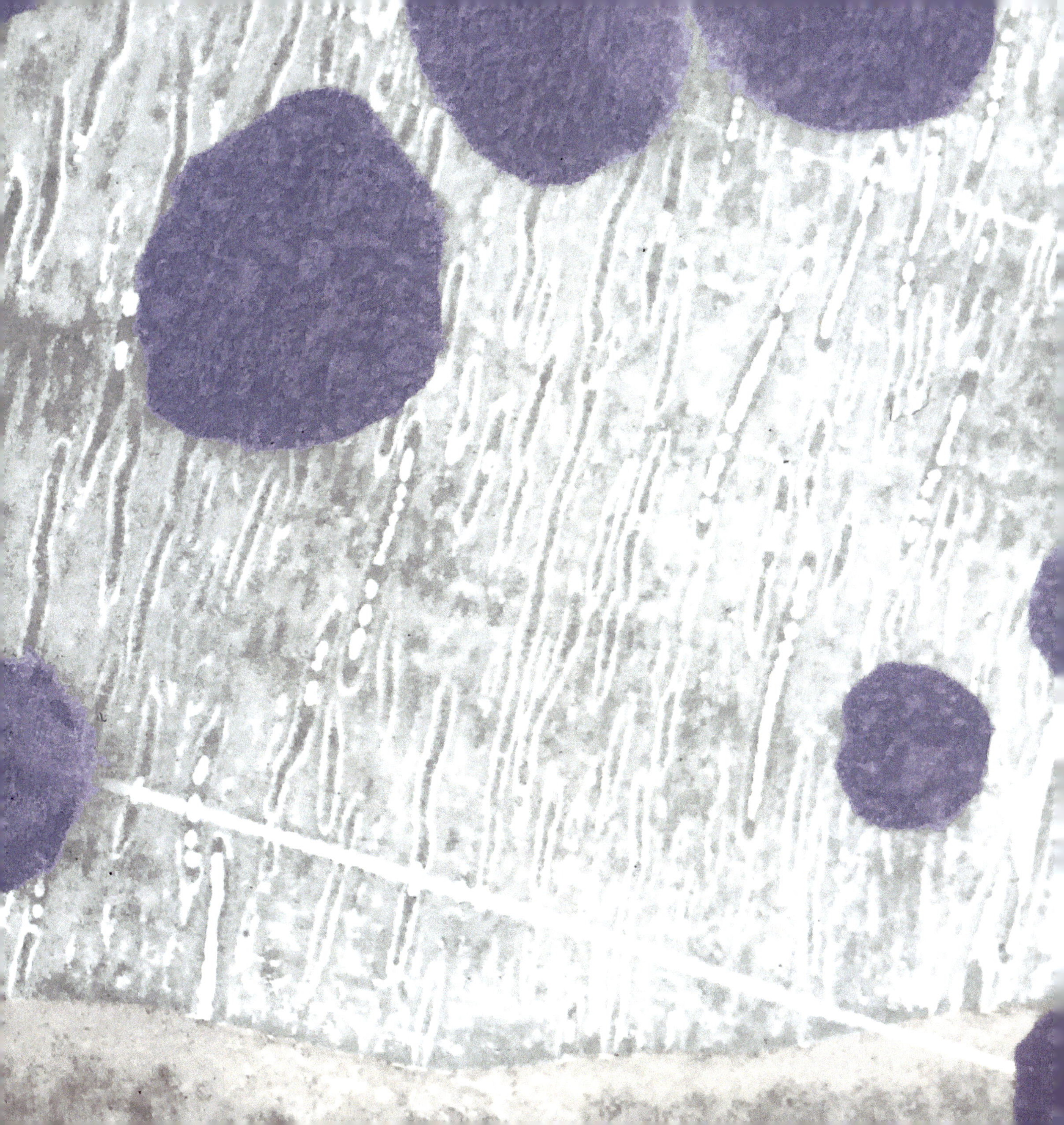

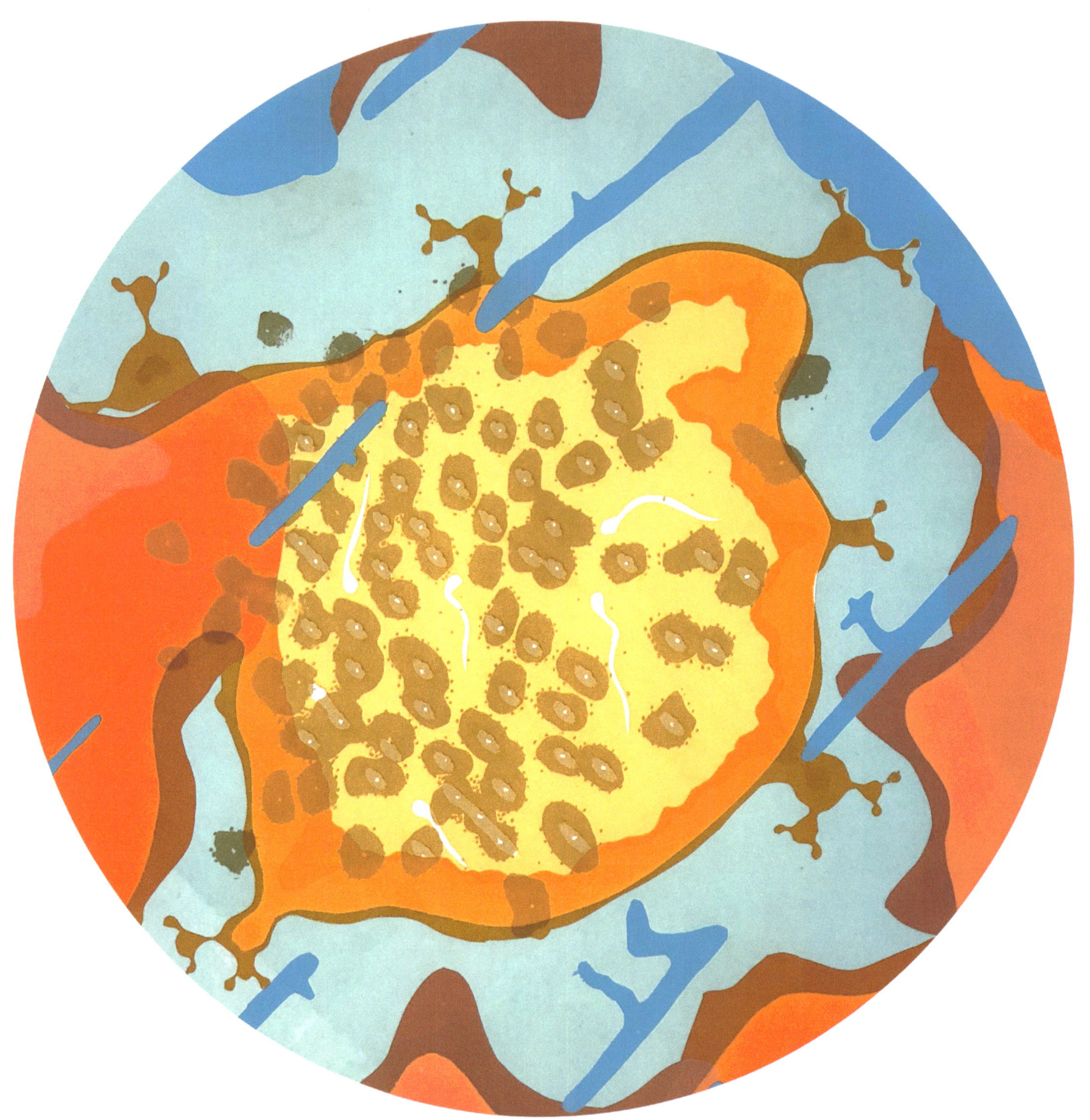

Existence conceived
a boundary
to encompass Entropy.

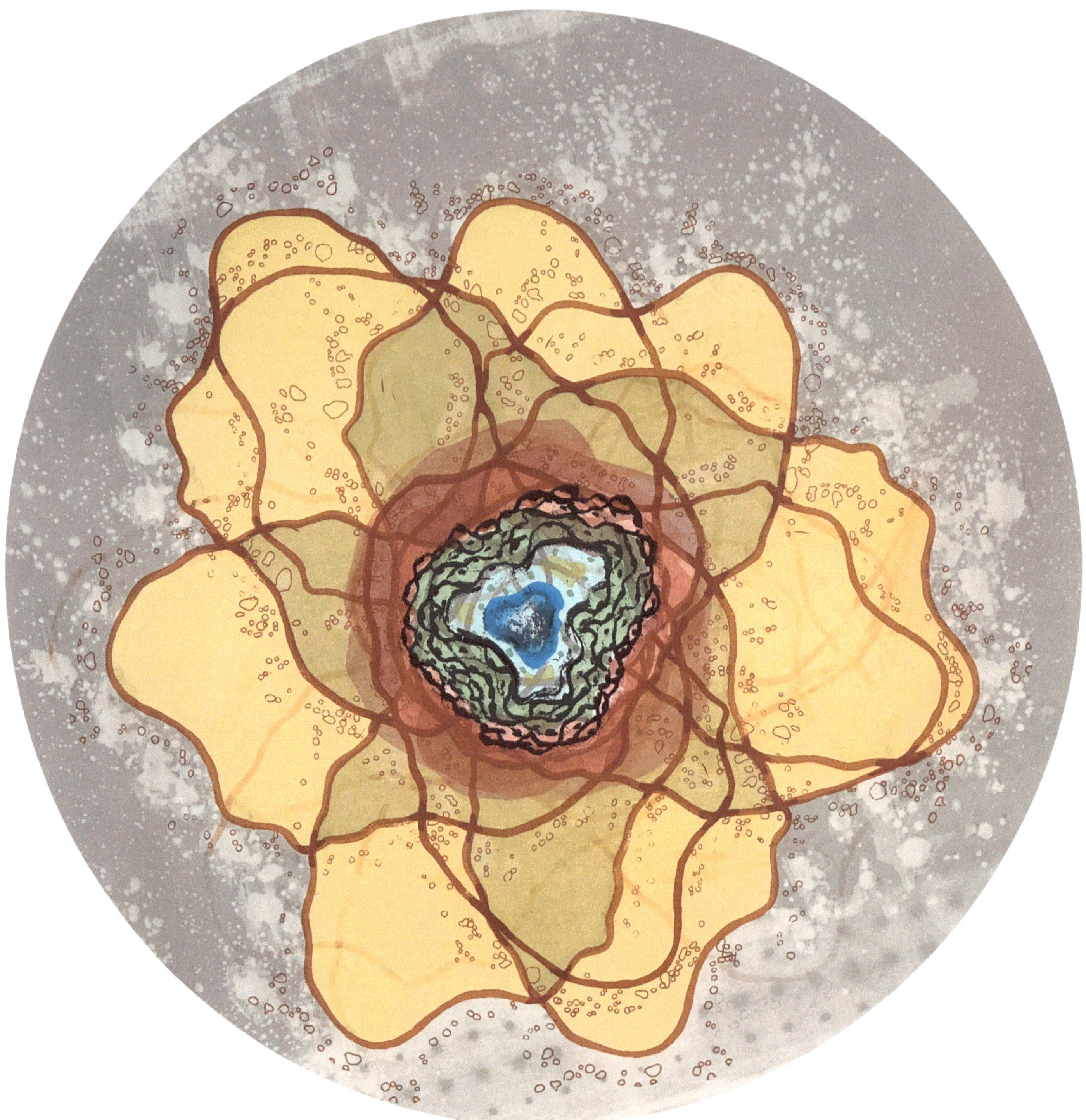

In a bang,
Existence reached out,
grasping Entropy
and change came into existence.

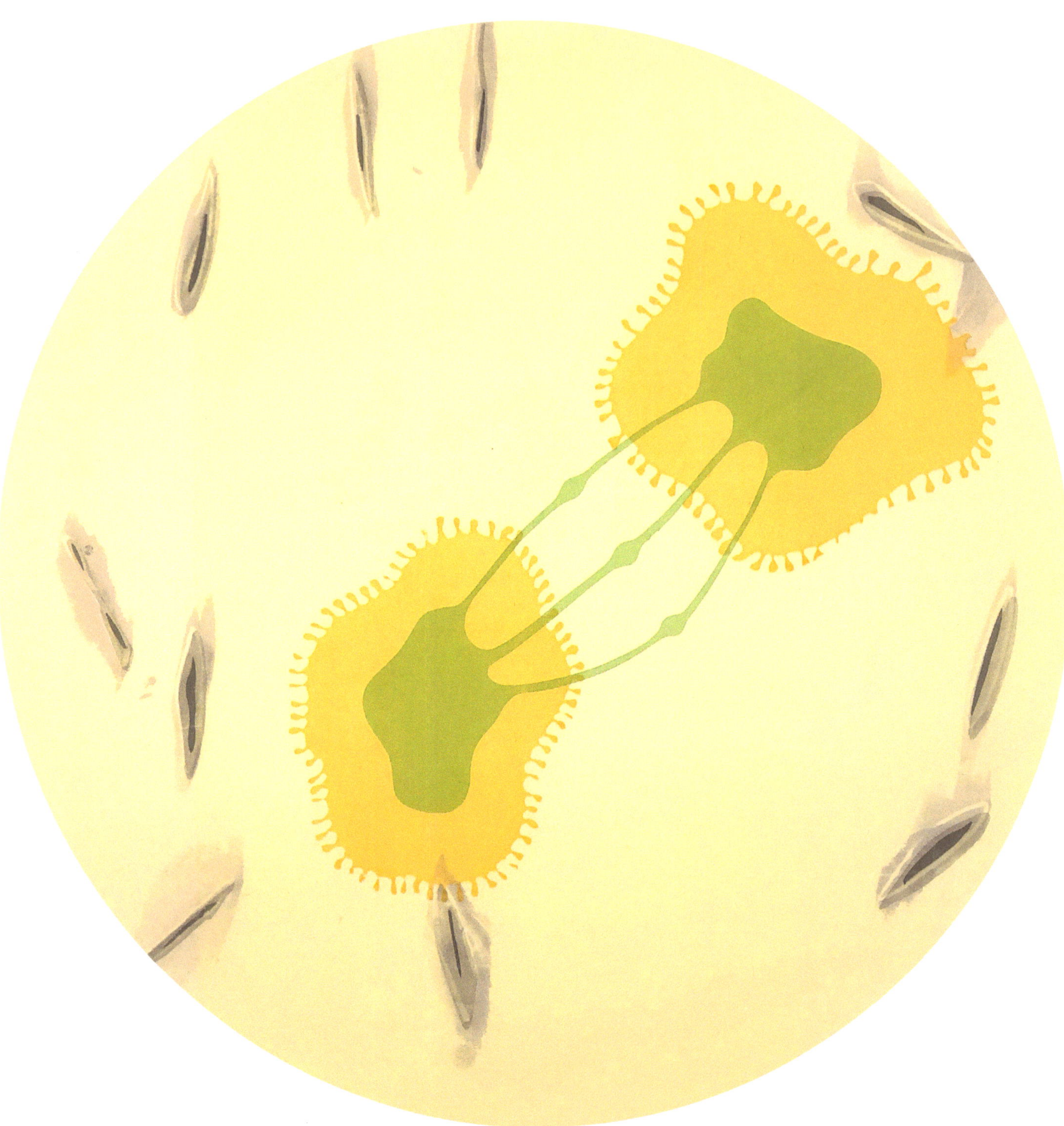

Here and there became;
so space became.

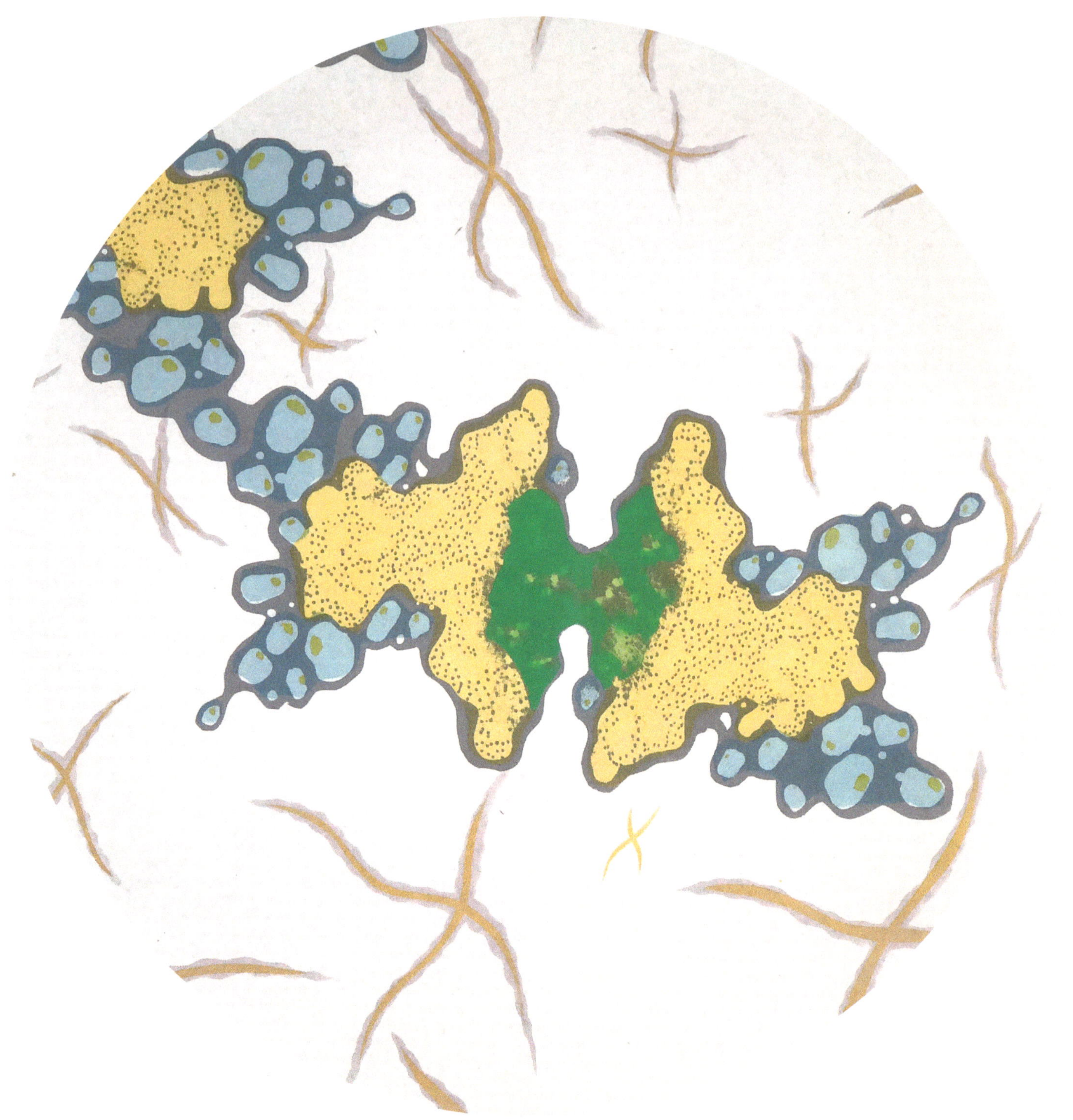

Before and after became;
so time became.

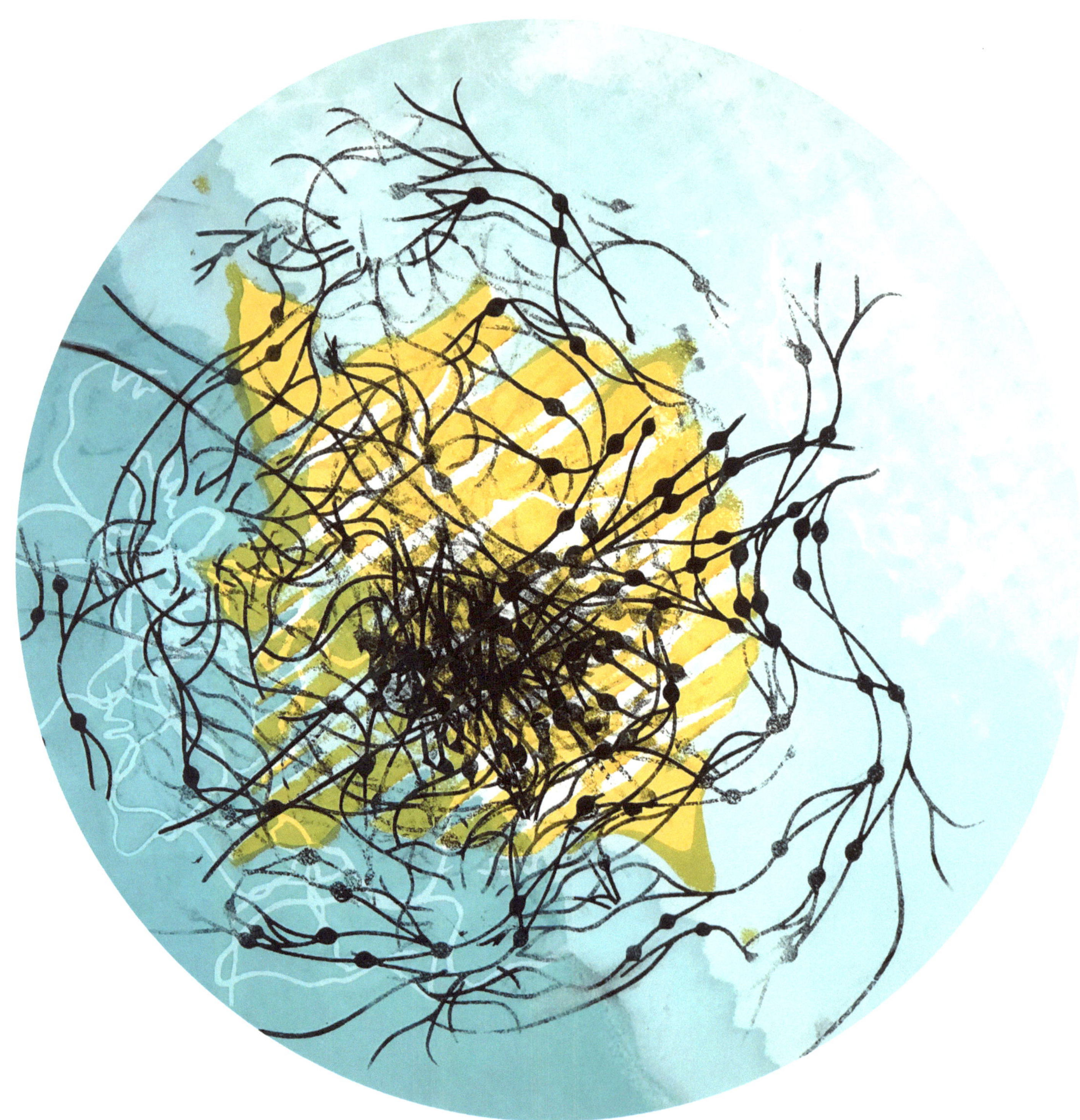

With space and time created,
Entropy was no longer zero
but Existence was no longer unified.

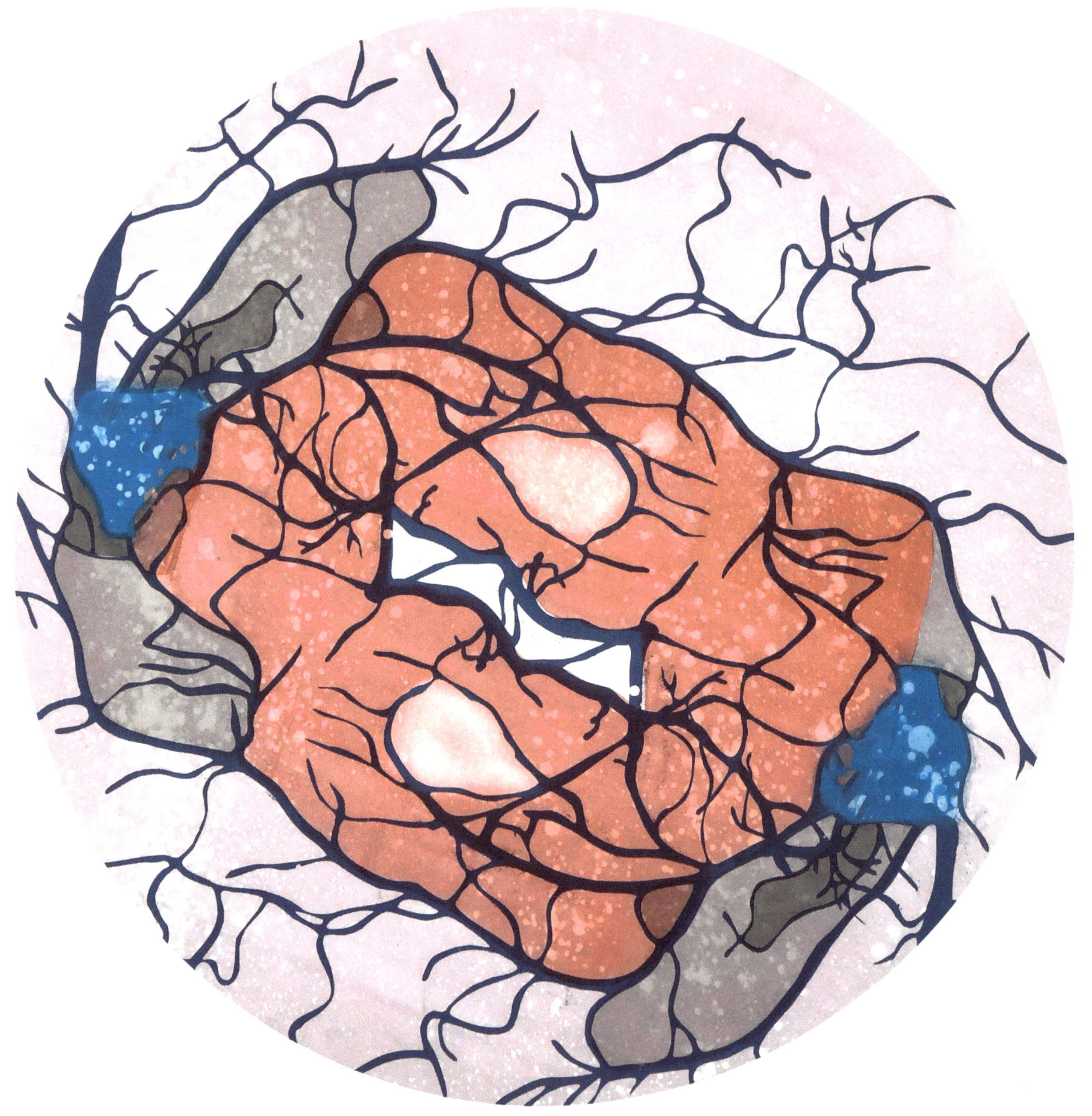

Symmetries broke,
forces and particles coalesced.

Existence clutched
at the dispersing patterns
that were once part of its unity.

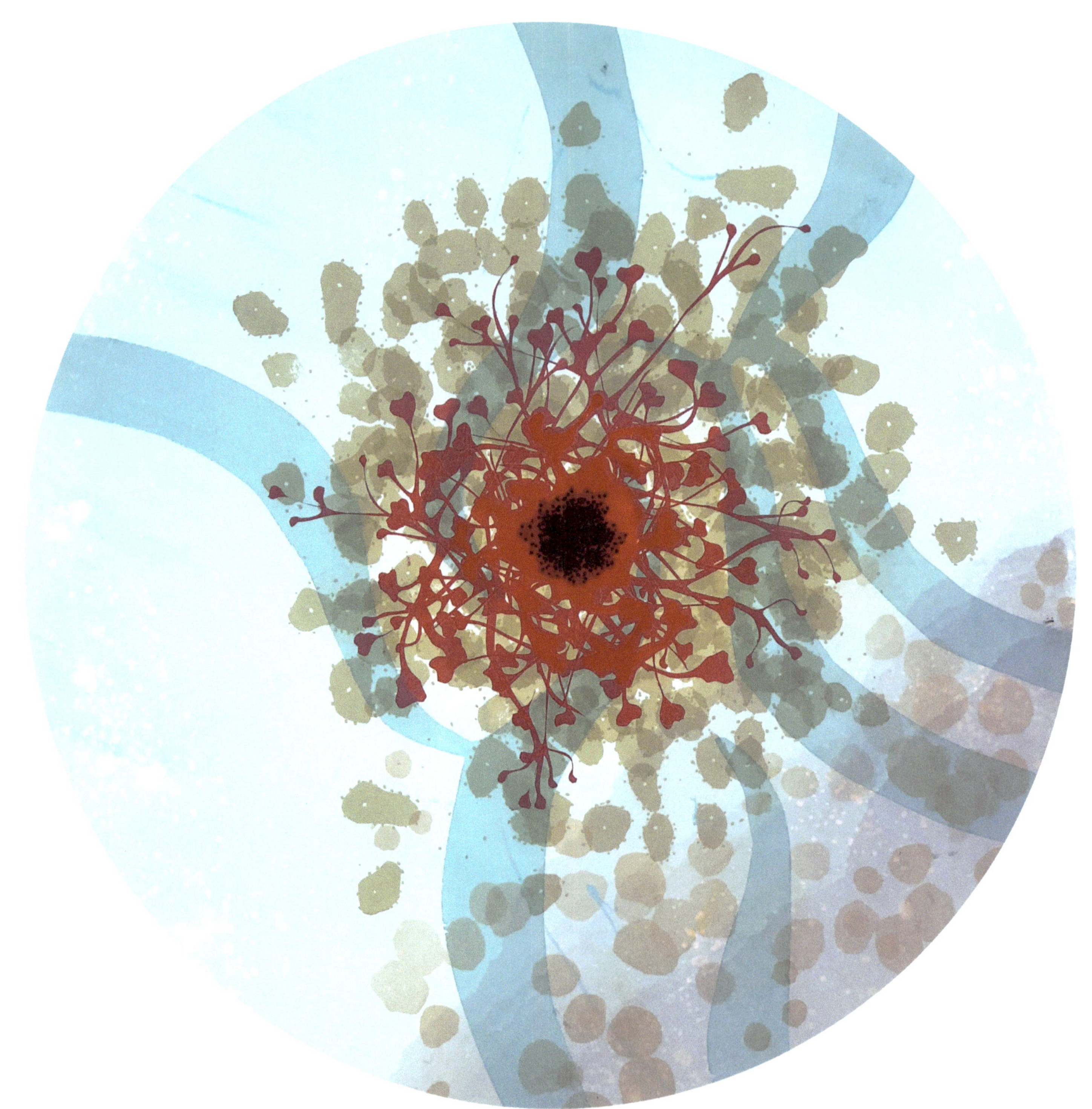

Existence created suns
to work at pressing
the new components of the universe
back toward their original form.

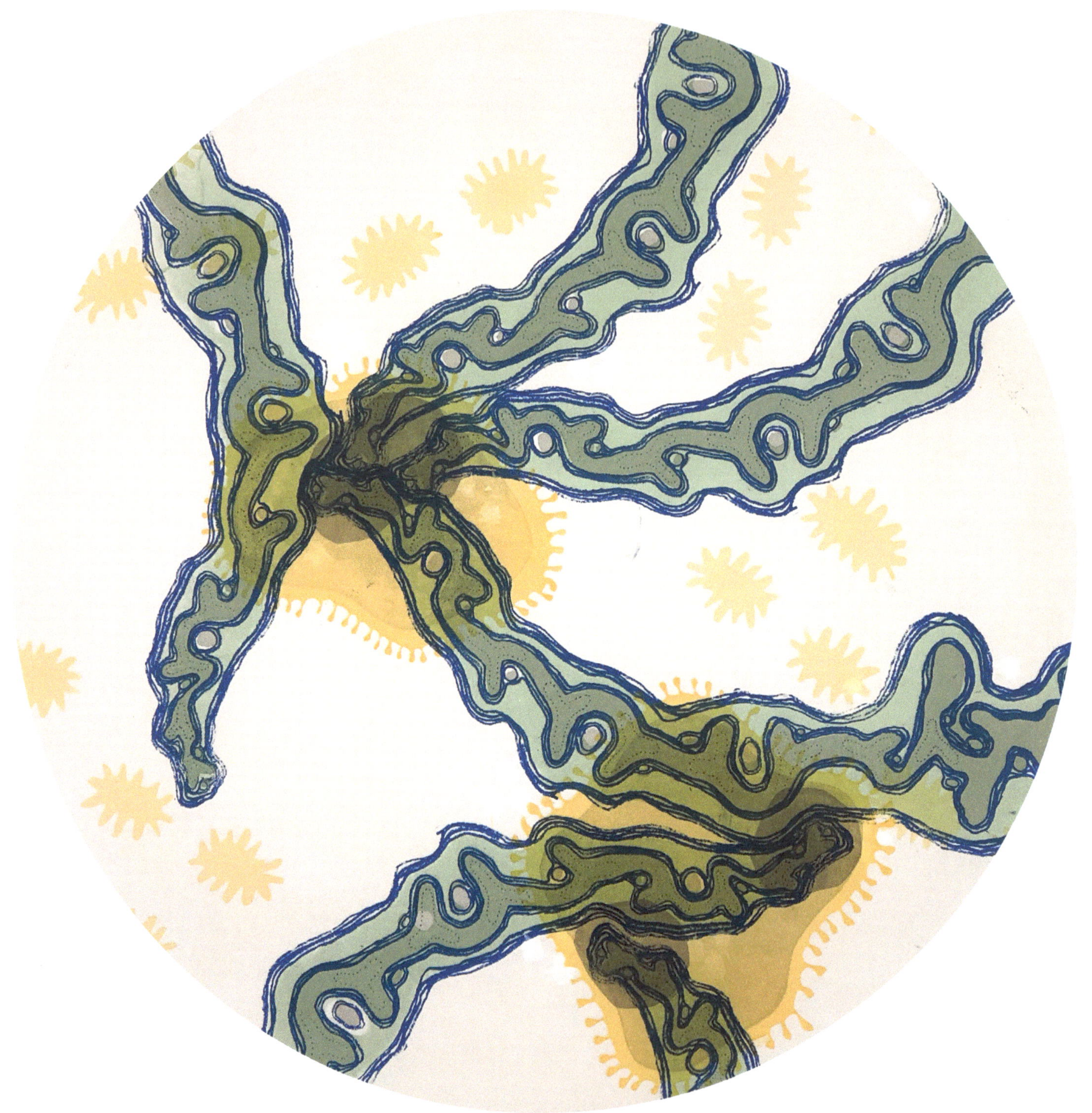

But always Entropy took its toll
and reunification fell short.

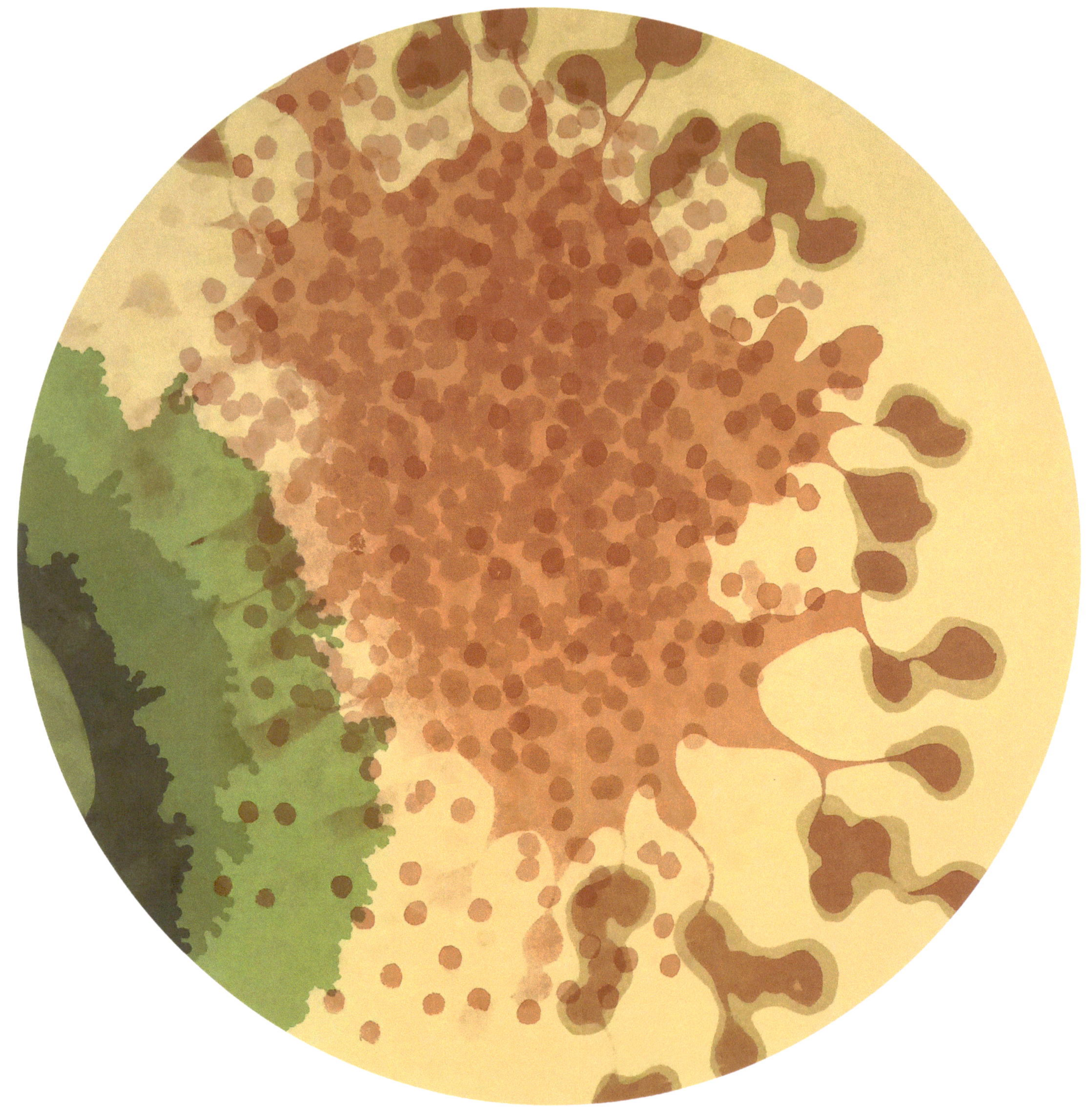

Everywhere,
at every scale
Existence worked to minimize Entropy
and return to its complete
—if imperfect—
unity.

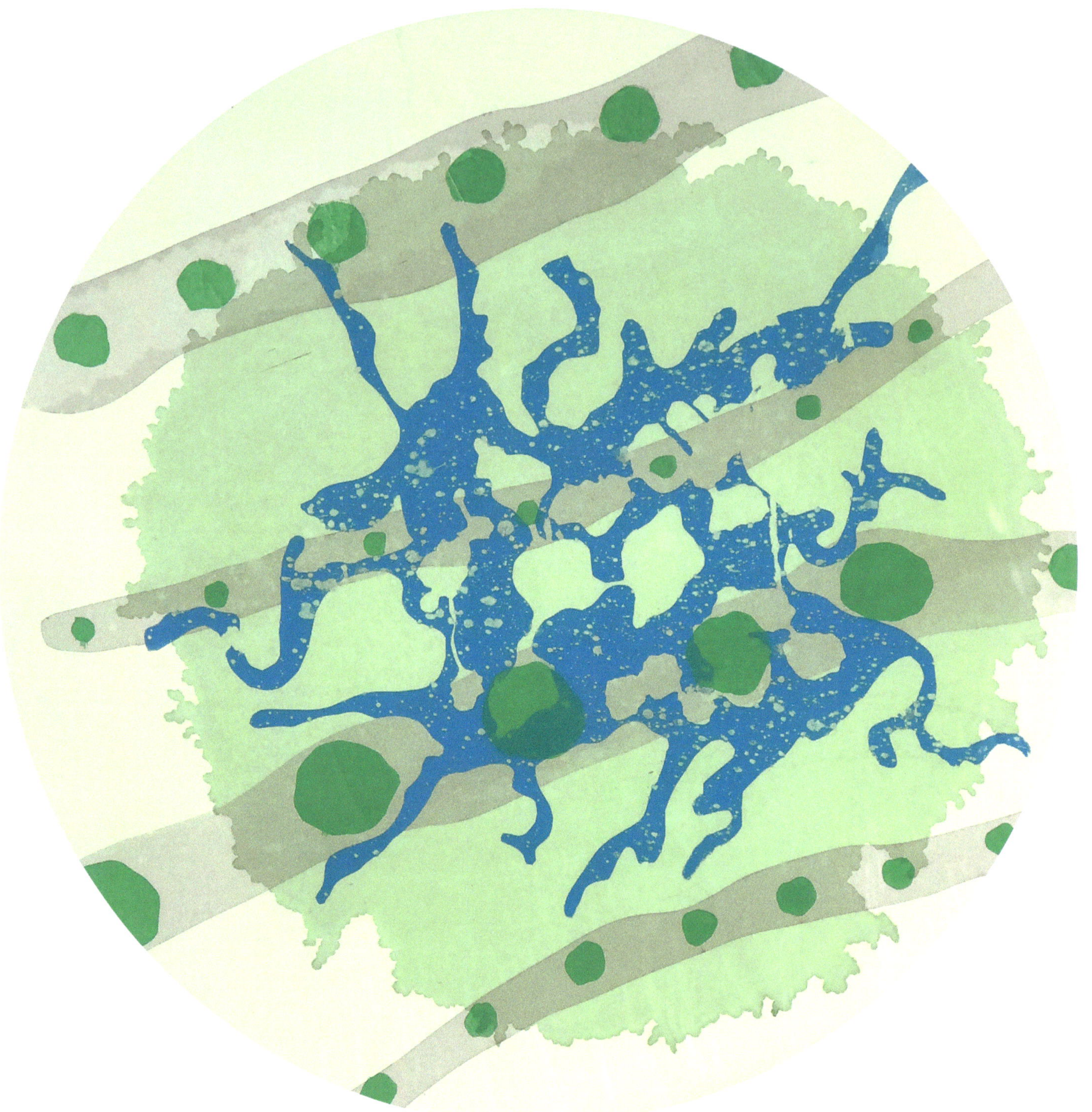

But every effort
was no more than an exchange;
a pattern here
for increased entropy
elsewhere.

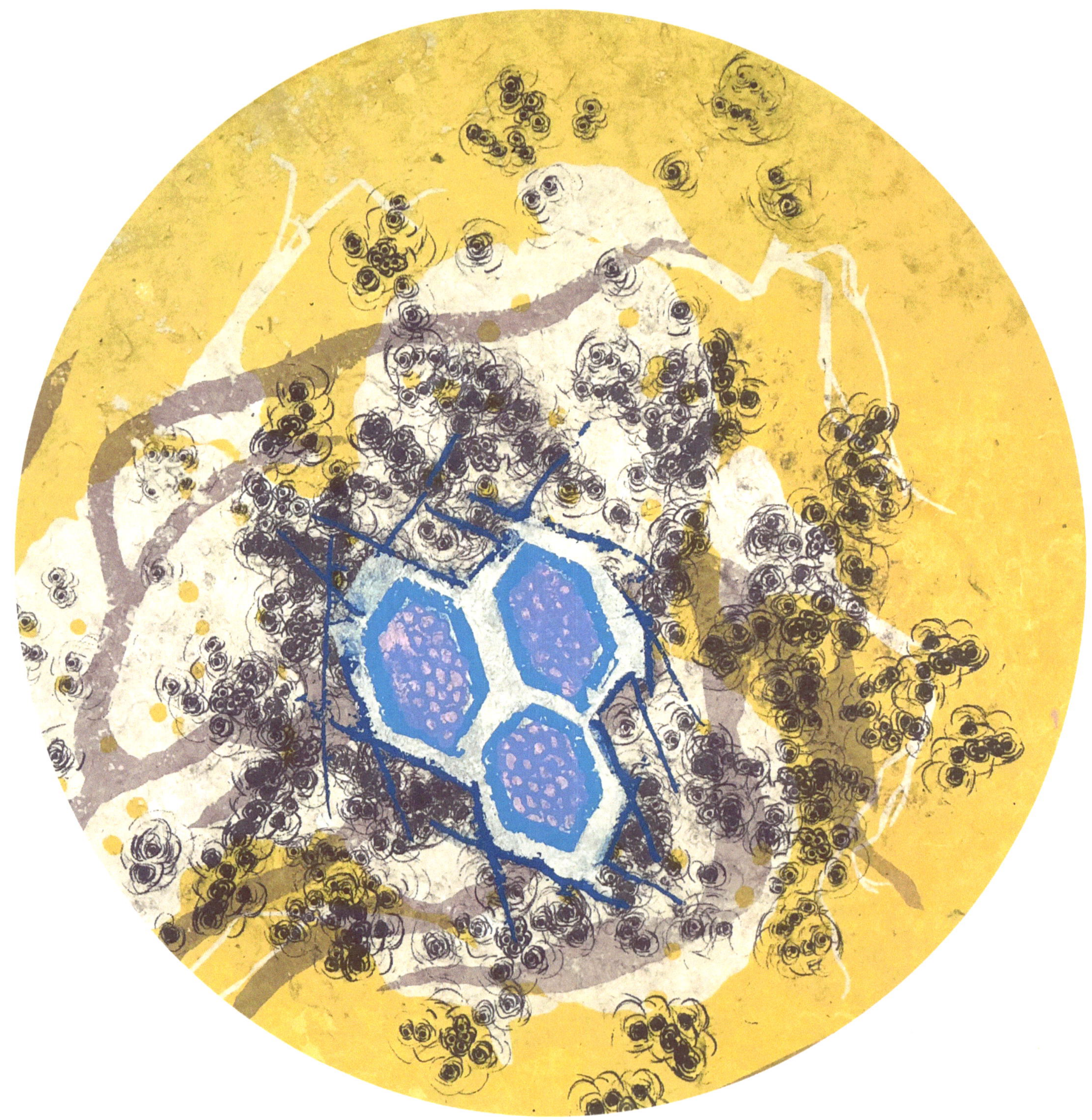

But each exchange
also increased the raw materials
available for new patterns
and each pattern
was another opportunity
to press Entropy back toward zero.

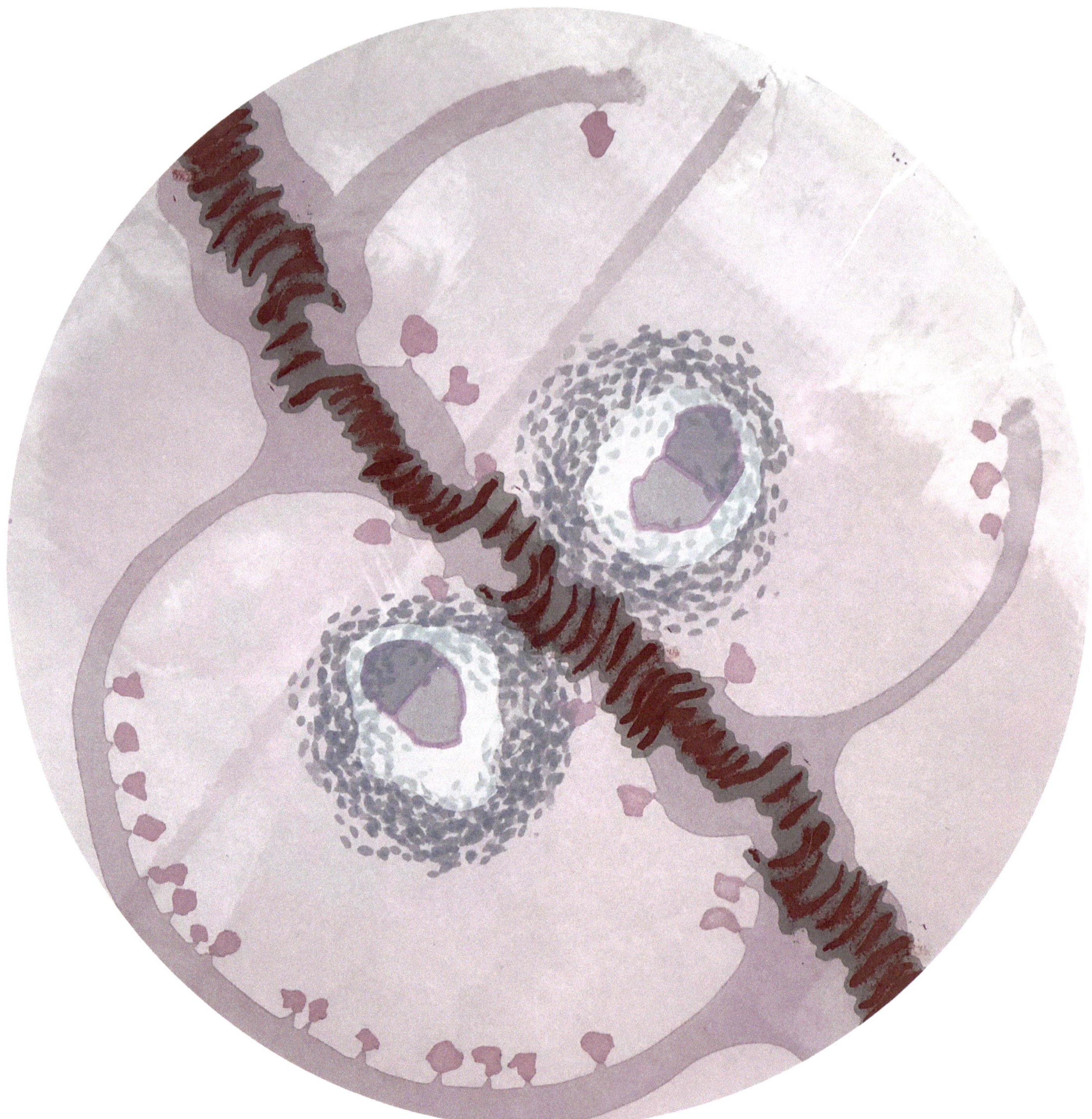

Soon Existence devised life,
an independent and self replicating pattern.

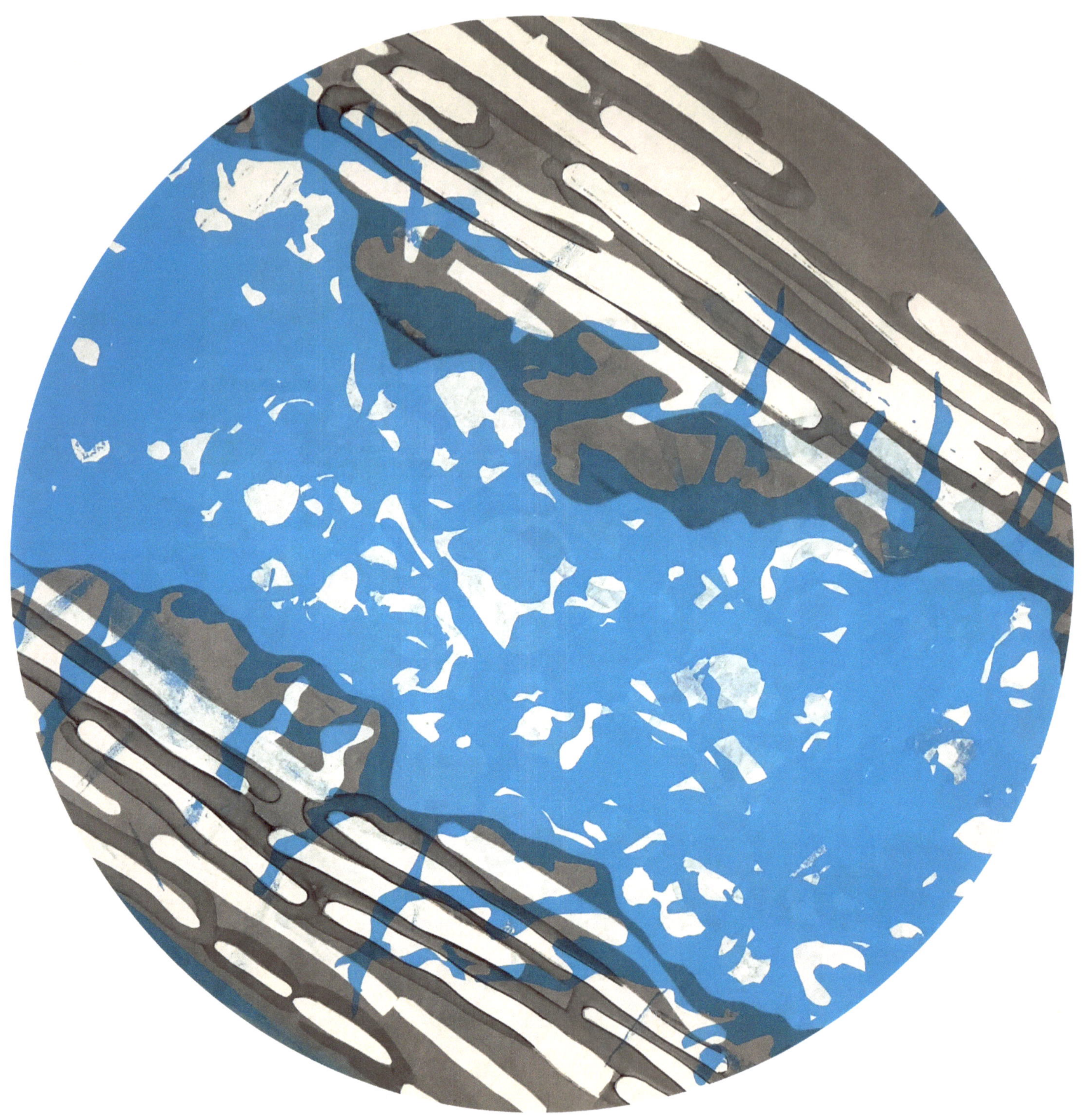

It launched life
across the expanding field
of space and time
to carry on the effort
to reclaim its lost glory.

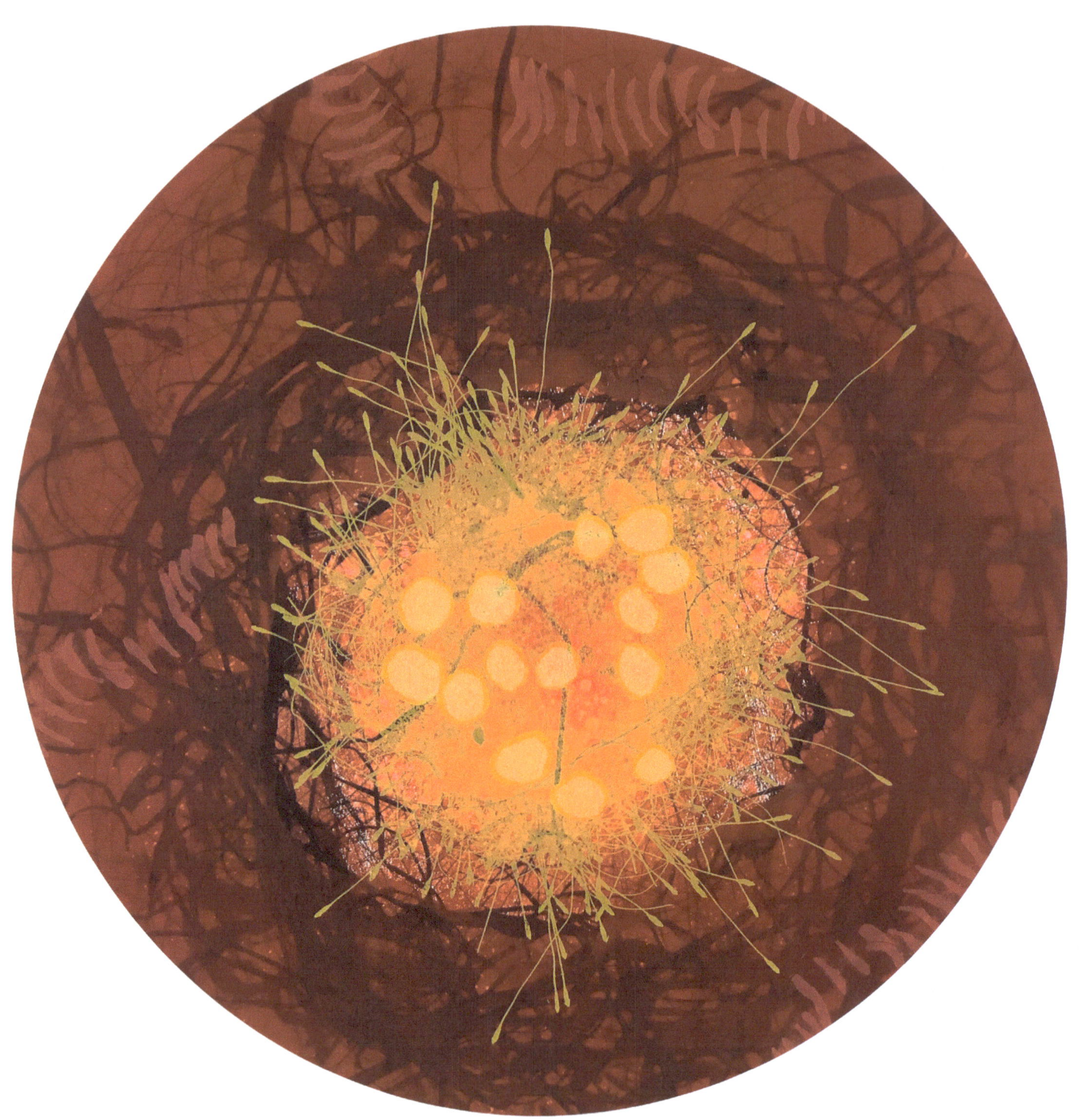

Is it possible
Existence still believed
it could return
to its original form?

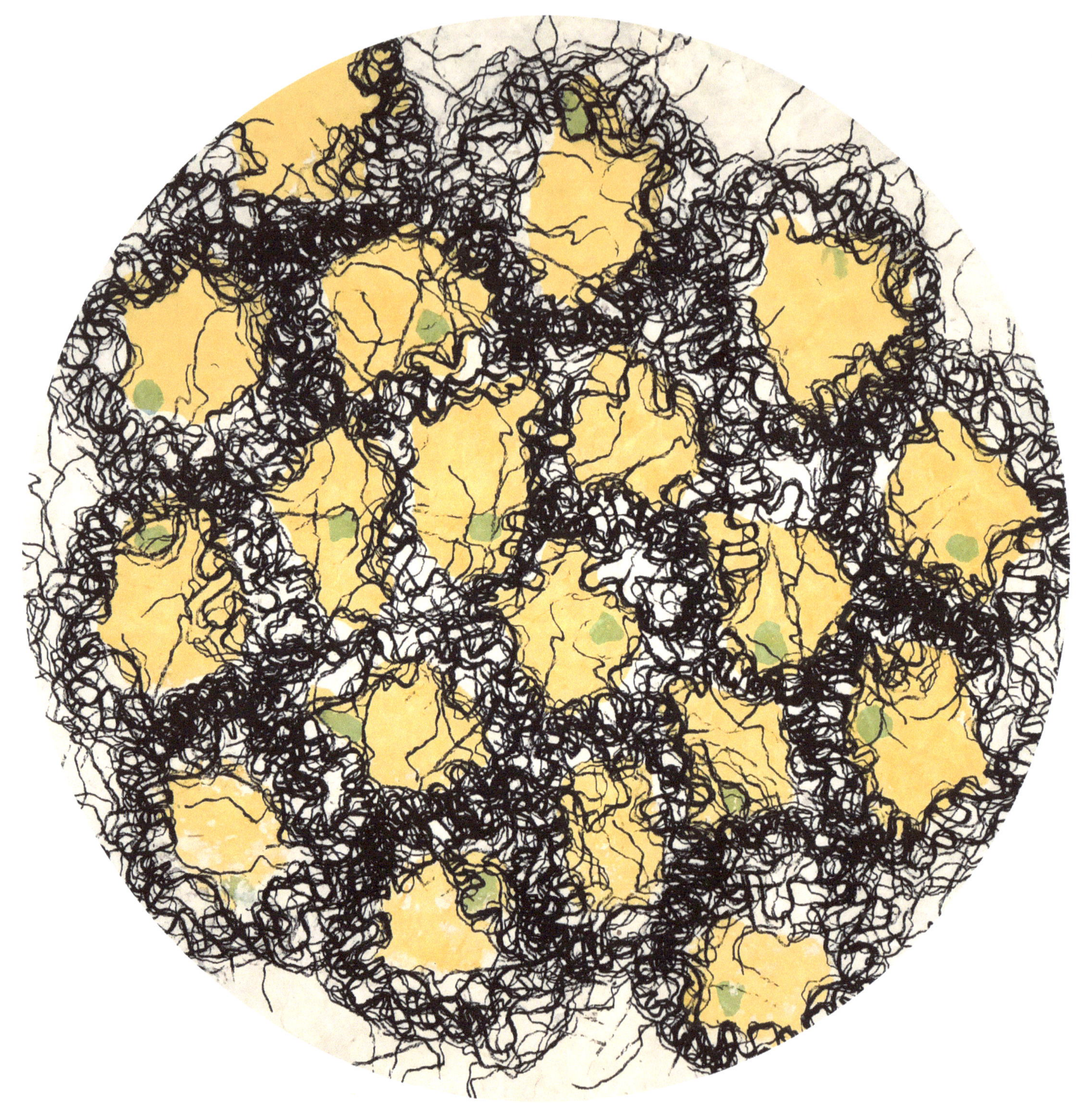

Did it believe life,
independent and inventive,
could find a path back?

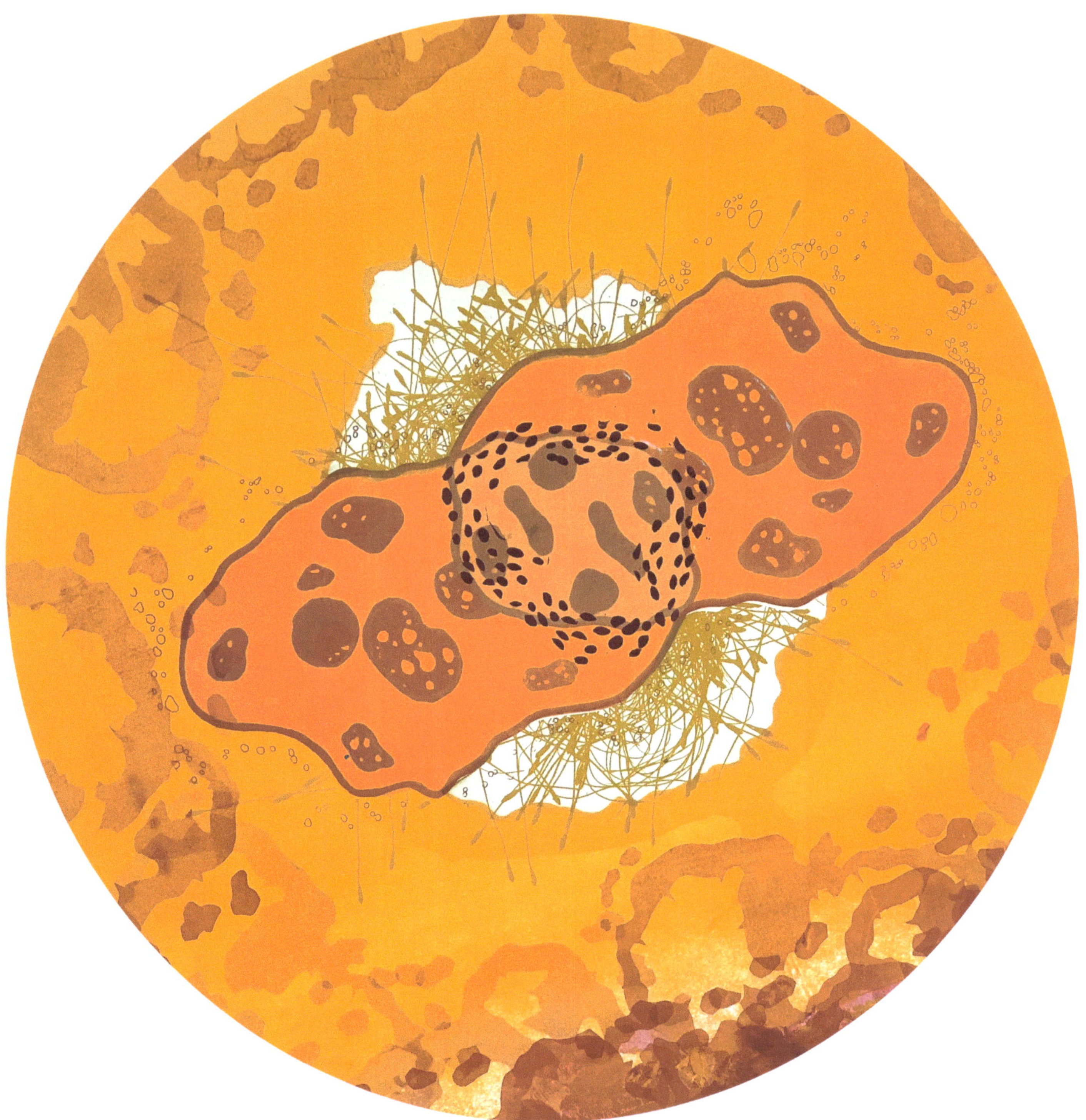

And should our allegiance be
to a stable unity
or to ecstatic change?

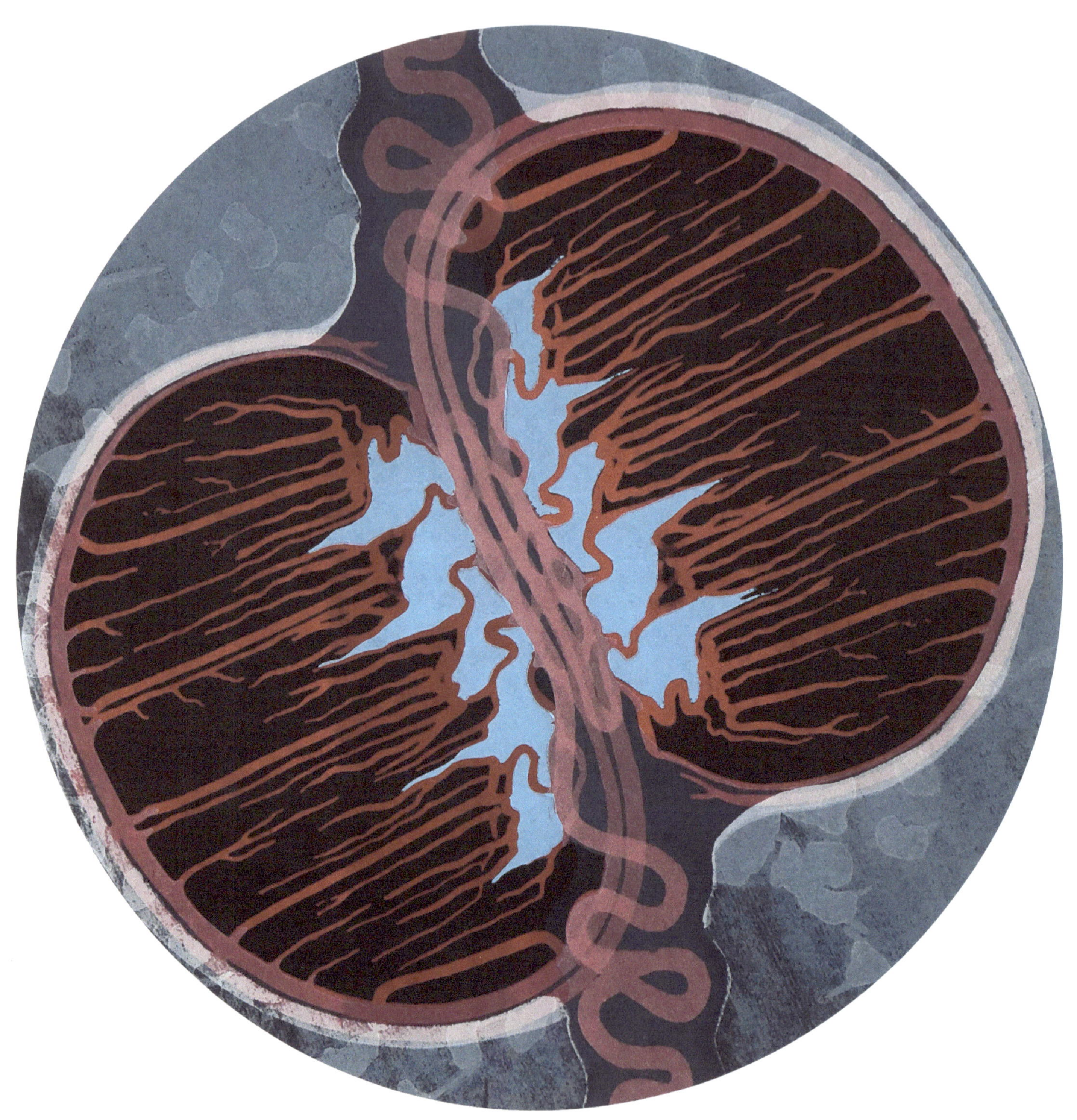

Existence would have us pursue
directed change toward an ideal.
But we might also
embrace change
as the source of
and the nature of
reality.

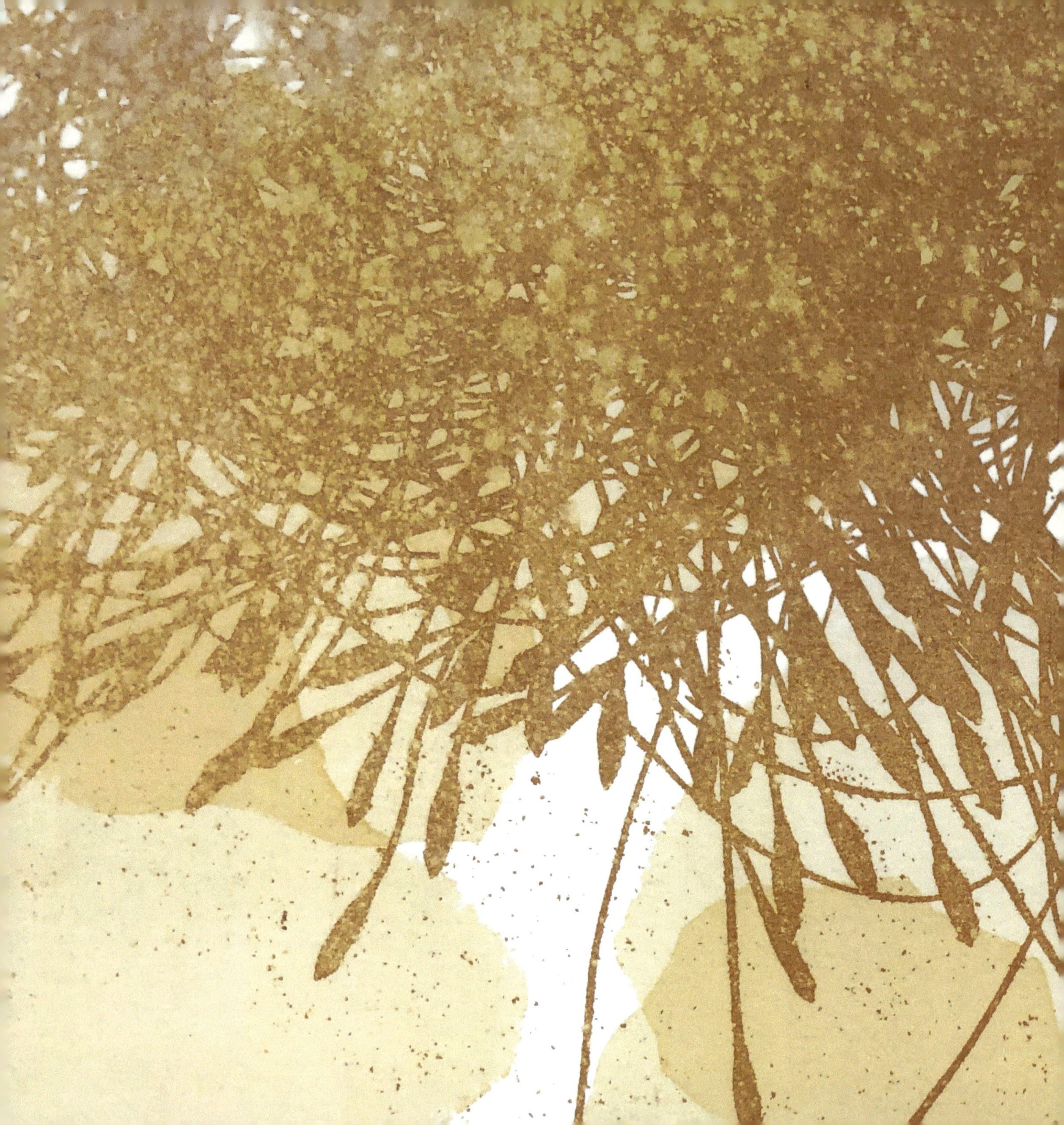

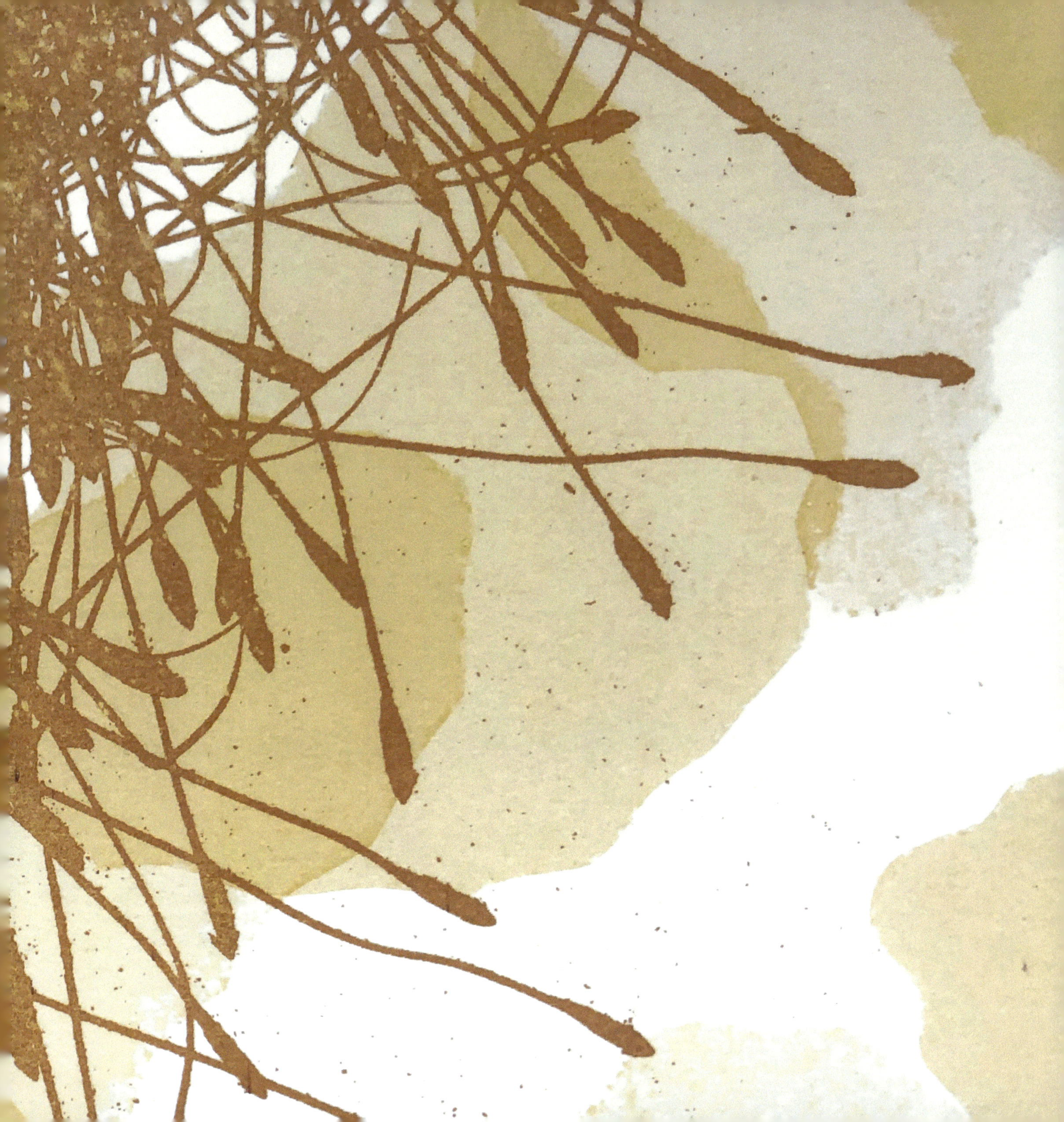

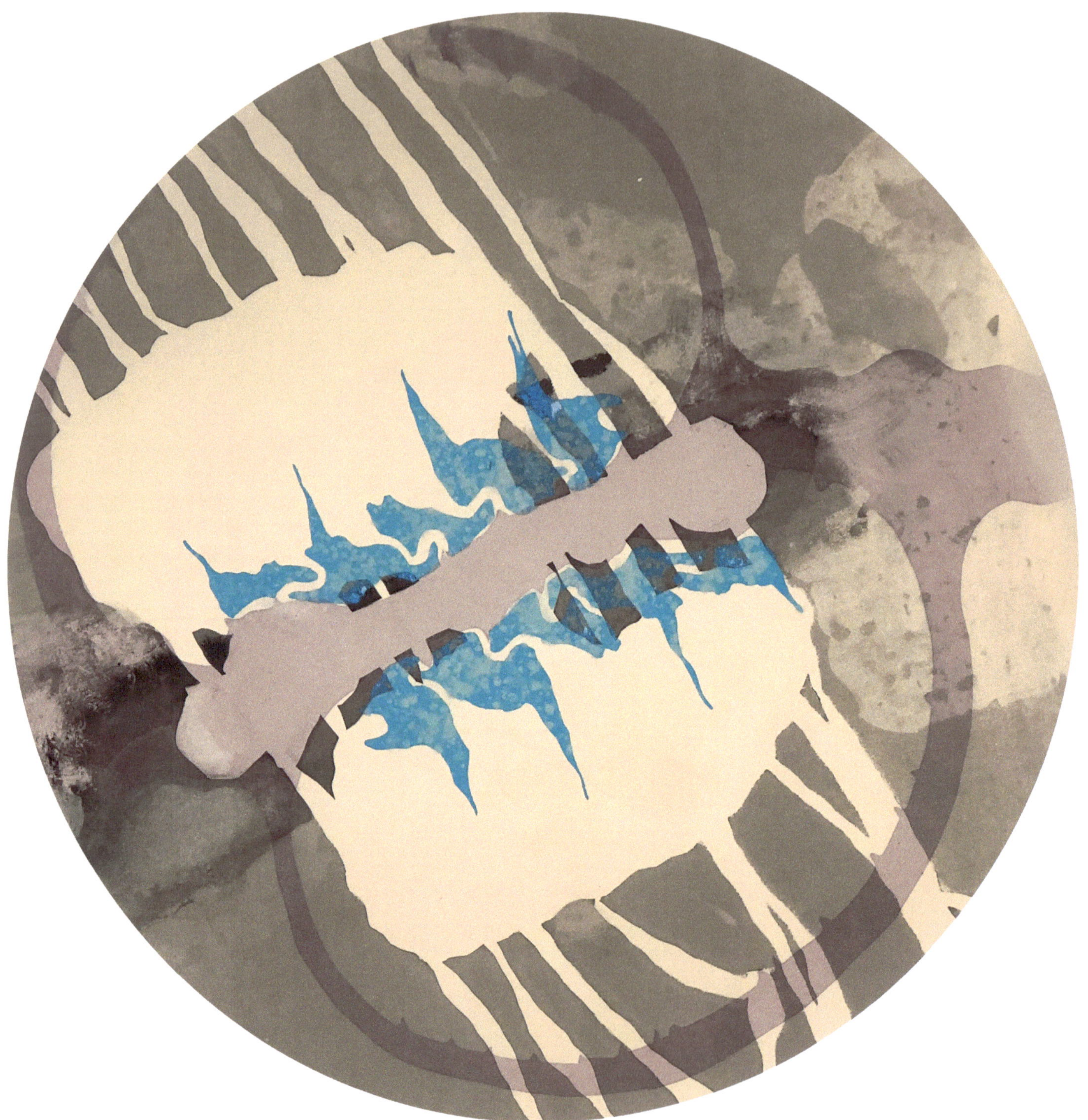

We are not the children of unity,
but of the effort to organize difference.

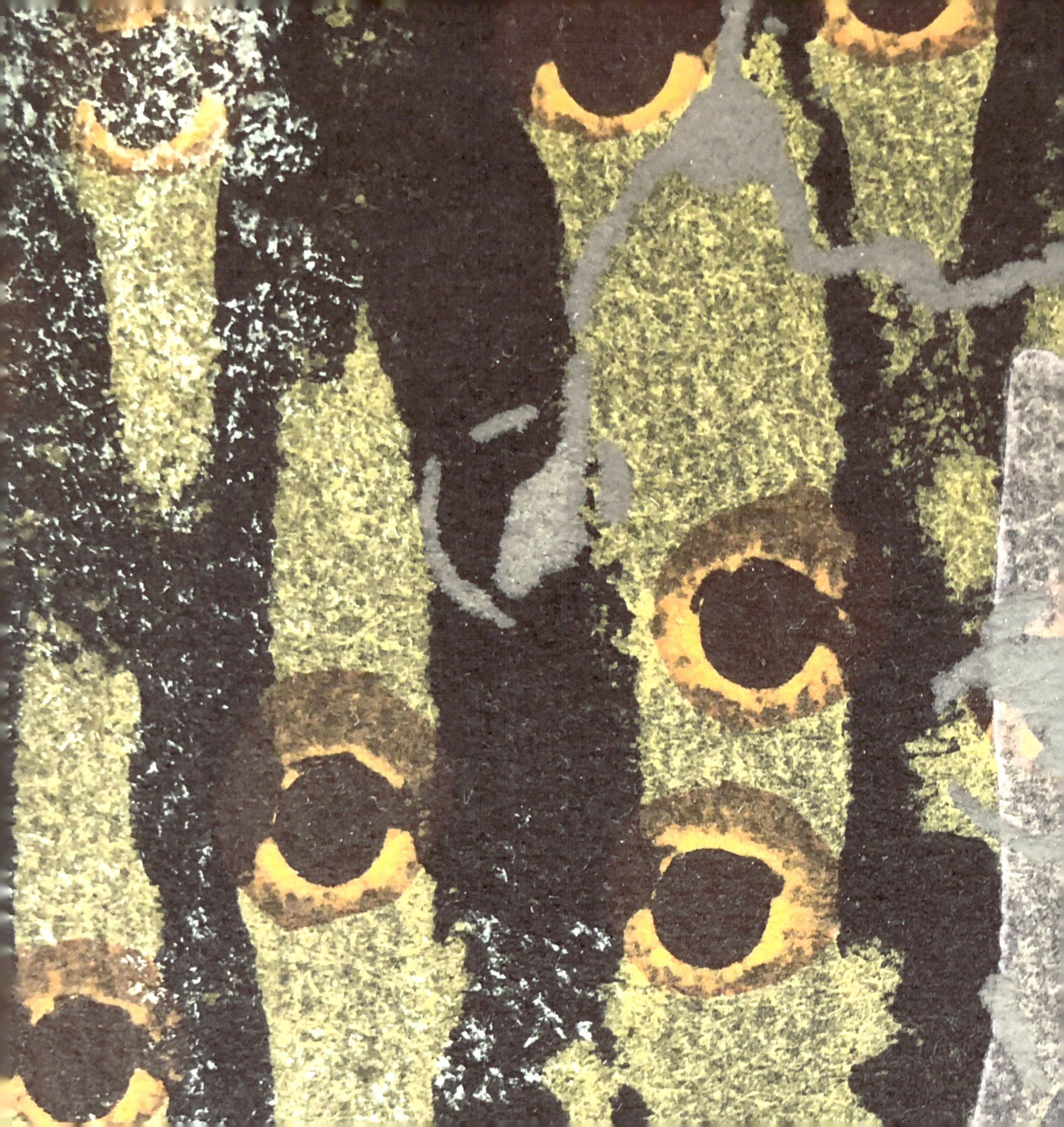

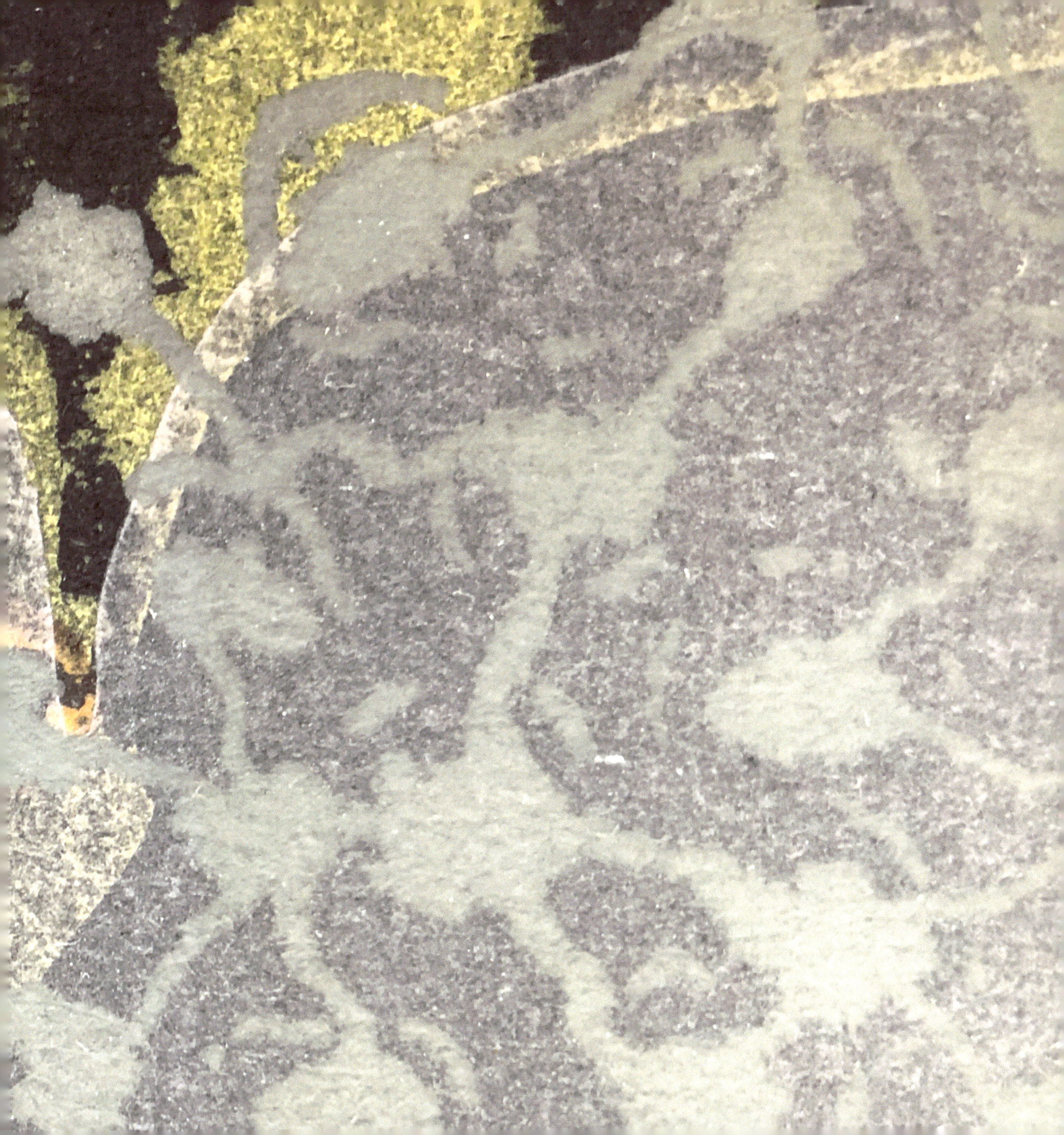

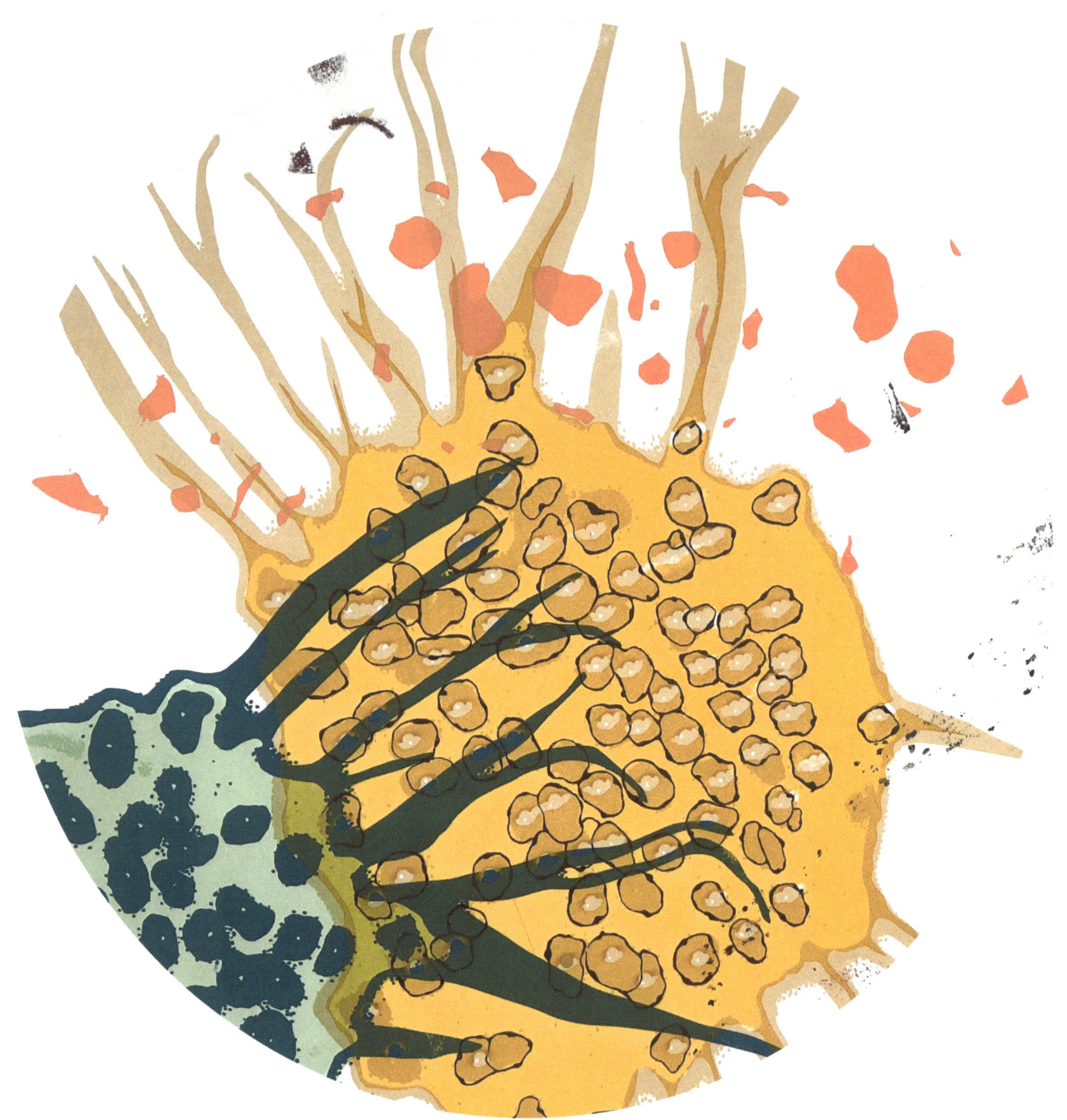

We exist
because in the attempt for perfection,
perfection was lost
while everything else was gained.

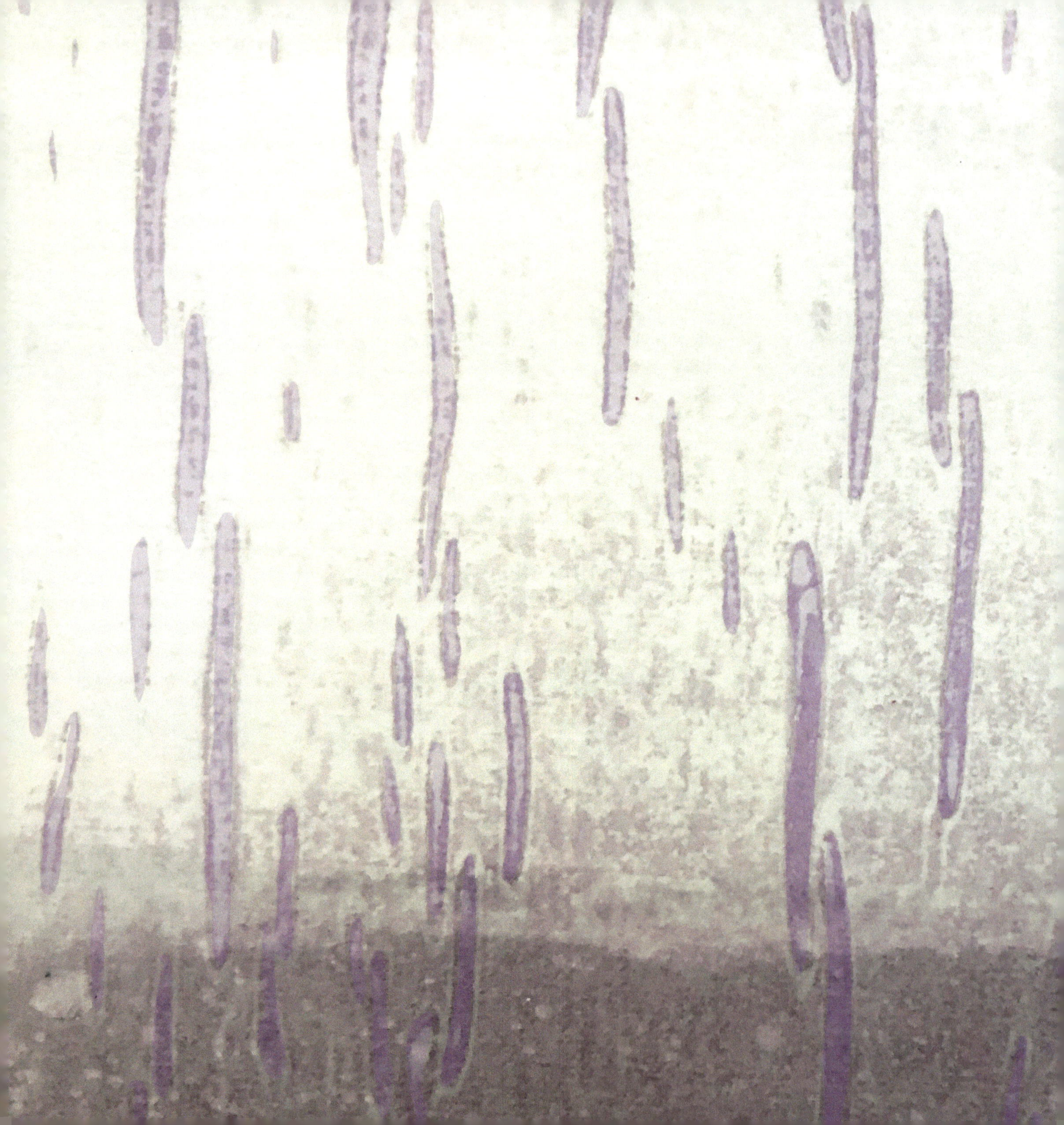

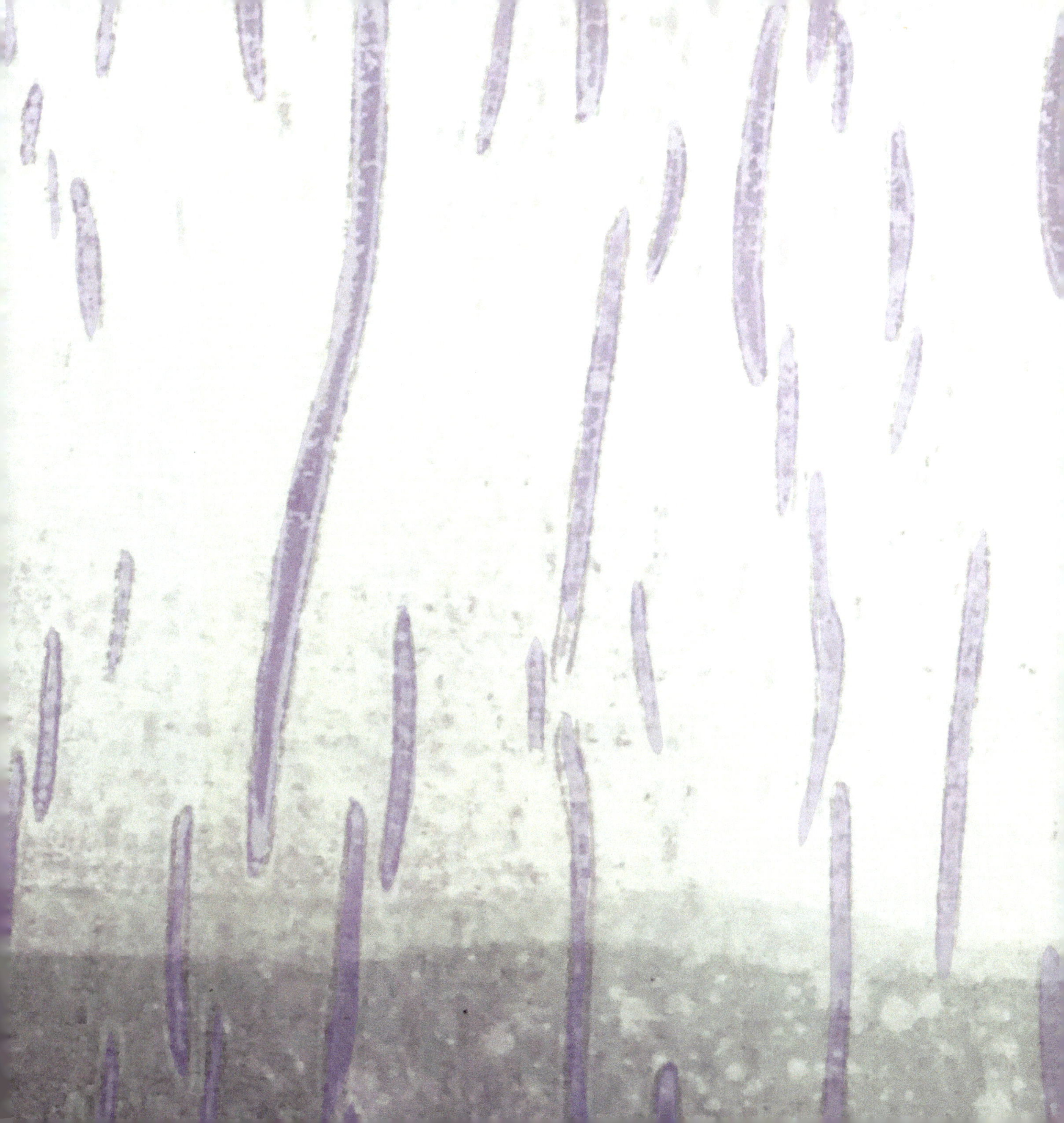

Don Drake wrangles his serial obsessions at Dreaming Mind Studio. He may be found there writing, making book art, coding, landscaping, or woodworking. Collaborations are his preferred form of social interaction and so he will often be found engaging in one of these actives with a fellow artist.

dreamingmind.com

Kent Manske creates images and symbols to inquire, process, manage, convey and assign meaning to ideas about human existence. He is a co-founder of PreNeo Press whose activities include art interventions, curating exhibitions and publishing.
His art life includes gardening, cultivating food, being in nature and exploring culture.

kentmanske.preneo.org

and change became

Published by Hunger Button Books
hungerbutton.org

ISBN: 978-1-936083-12-1

Design & Production: Don Drake & Kent Manske
Printed in the United States of America